Electronic Servicing and Repairs

To Joyce, Samantha and Victoria

Electronic Servicing and Repairs

Third Edition

TREVOR LINSLEY
Senior Lecturer
Blackpool and The Fylde College

Newnes
OXFORD • AUCKLAND • BOSTON • JOHANNESBURG • MELBOURNE • NEW DELHI

Newnes
An imprint of Butterworth-Heinemann
Linacre House, Jordan Hill, Oxford OX2 8DP
225 Wildwood Avenue, Woburn, MA 01801-2041
A division of Reed Educational and Professional Publishing Ltd

A member of the Reed Elsevier plc group

First published as *Electronics for Electricians* by Edward Arnold 1990
Second edition published as *Electronics for Electricians and Service Engineers* 1993
Third edition published by Butterworth-Heinemann 2000

© Trevor Linsley 2000

All rights reserved. No part of this publication may be reproduced in
any material form (including photocopying or storing in any medium by
electronic means and whether or not transiently or incidentally to some
other use of this publication) without the written permission of the
copyright holder except in accordance with the provisions of the Copyright,
Designs and Patents Act 1988 or under the terms of a licence issued by the
Copyright Licensing Agency Ltd, 90 Tottenham Court Road, London,
England W1P 0LP. Applications for the copyright holder's written
permission to reproduce any part of this publication should be addressed
to the publisher

British Library Cataloguing in Publication Data
A catalogue record for this book is available from the British Library

ISBN 0 7506 5053 2

Typeset by Phoenix Photosetting, Chatham, Kent
Printed and bound in Great Britain by MPG Books Ltd, Bodmin, Cornwall

FOR EVERY TITLE THAT WE PUBLISH, BUTTERWORTH-HEINEMANN
WILL PAY FOR BTCV TO PLANT AND CARE FOR A TREE.

CONTENTS

Preface to the third edition vii

1 **Health and safety** 1
The Health and Safety at Work Act 1974 – The Control of Substances Hazardous to Health Regulations 1988 (COSSH) – Personal Protective Equipment (PPE) – The Electricity at Work Regulations 1989 (EWR) – British and European standard marks – Electrical safety in workshops and laboratories – Risk and hazard – Safety signs – First aid at work – Accident reports – Fire control – Manual handling – Tidiness – Personal Hygiene – Ergonomics – Personal awareness – VDU operation hazards – Exercises

2 **Electronic component recognition** 30
Electronic circuit symbols – Abbreviations used in electronics – The standard colour code – 5 band metal film resistors – Capacitors – Coupling and decoupling capacitors – Inductors and transformers – Switches – Electromagnetic relay – Overcurrent protection – Supplies – Packaging electronic components – Obtaining information and components

3 **Electronic circuit assembly** 45
Statutory regulations – Safety precautions – Hand tools – Soldering irons – Soldering gun – Soldering – Component assembly and soldering – Desoldering – Removing faulty transistors – Removing faulty integrated circuits (chips) – Wire wrapping – Breadboards or prototype – Interconnection methods – Fault finding

4 **Semiconductor devices** 58
Semiconductor materials – Semiconductor diode – Light-dependent resistor (LDR) – Thermistor – Transistors – Integrated circuits – The thyristor or silicon-controlled rectifier (SCR)

5 **Electronic circuits in action** 71
Voltage divider – Rectification of a.c. – Smoothing – Diode and capacitor ratings – Stabilised power supplies – Fixed voltage series regulators – Flywheel diode – Three-phase power control using thyristors – Amplifiers – Filters – Testing audio amplifiers – Waveforms – Harmonics – Signal modulation – Sawtooth waveform generator – Transistor switching using a capacitor

6 **Testing electronic circuits** 93
Test instruments – Analogue and digital displays – The cathode ray oscilloscope (CRO) – Signal generators – Power supply unit (PSU) – Mains electricity supply – Insulation tester – Portable appliance testing (PAT)

7 **Digital electronics** 104
The AND logic gate – The OR gate – The exclusive -OR gate – The NOT gate – The NOR gate – The NAND gate – Buffers – Logic networks – Logic families – Comparison of TTL and CMOS – Working with logic – British standard symbols – Exercises

8 Electronic circuit theory 113

Units – Basic circuit theory – Resistivity (symbol ρ – the Greek letter 'rho') – The three effects of an electric current – Electrostatics – Capacitors – Resistors – Power and energy – Alternating current theory – Resistance and reactance in an a.c. circuit – Resistance, inductance and capacitance in an a.c. circuit – Power and power factor – Resistance, inductance and capacitance in series – Series resonance – Magnetism and motors – Transformers – Mechanics – Conservation of energy – Sound – Light – Reflection and refraction – Lenses – Colour – Exercises

9 Electronic systems 149

Open loop control – Closed loop control – Negative feedback – Positive feedback – Transducers – Security systems – Space heating control – AM transmitter – AM receiver – FM transmitter – FM receiver – Cathode ray oscilloscope or CRO – Tape recorder – Digital clock – Computing systems – Drawings and diagrams

10 Communication systems 163

Simple communications systems – Simple telephone circuit – The modern telephone system – Optical fibres – System X – Microwave telephone links – Satellite telephone links – Mobile telephones – Telephone at home – Radio transmission – Geostationary satellite communications – Satellite television – Satellite dish installation – Fine tuning for maximum signal strength – Regulations – Cable television – Computer supplies – Computer networks – Local area networks of computers (LANs)

11 Security systems 176

Security lighting – PIR detectors – Intruder alarm systems – Intruder alarm sounders – Design considerations – Closed circuit television (CCTV) – Fire security systems

12 Sensors and transducers 186

Transducers and process control – Measurement of strain – Active and passive axis – Actual measurements – Measuring the output voltage – Bonding the strain gauge – Measurement of pressure – Piezoelectric pressure transducers – Bourdon tube pressure transducer – Microphone pressure transducer – Measurement of temperature – Thermocouple temperature transducer – Thermistor temperature transducer – Measurement of liquid level – Measurement of fluid flow – Measurement of speed of rotation – Pulse counter – The Bourdon tube pressure gauge – Stepper motor – Industrial sensors – Opto electronics

Appendices 209

A Obtaining information and components
B Abbreviations, symbols and codes
C Greek symbols
D Battery information
E Small signal diodes
F Power diodes
G Zener diodes
H Transistors
I Voltage regulators
J Power control
K Comparison of British and American logic gate symbols
L Integrated circuit logic gates
M Thermocouple colour coding
N Strain gauges
O Health and Safety Executive (HSE) publications and information

Glossary 239

Solutions to exercises 248

Index 251

PREFACE TO THE THIRD EDITION

This book deals with many of the Health and Safety and Electronic Hardware topics in the City and Guilds Information Technology Schemes of Work numbers 7261 and 4242. It also covers the electronic topics of the City and Guilds 2351 Knowledge of Electrical Installation Engineering.

In addition to meeting the specific requirements of these courses, the book also provides a basic user friendly guide to electronics for enthusiasts and students on BTEC, SCOTVEC and other City and Guilds courses in Electrical and Electronic Engineering.

Electronics has today found its way into most industrial applications and consequently many craftsmen now find that a basic knowledge of electronics is essential to carry out their work efficiently. One of my aims in writing this book is to provide readers with a basic working knowledge of electronics which they can quickly apply to their own practical situation. The Appendices brings together some of the basic reference information which those new to electronics will find useful.

The treatment of electronics in this book is a non-mathematical one. However, for those who require a deeper understanding of electronic circuits, I have included chapter 8, Electrical Circuit Theory, which should perhaps be 'dipped into' on a 'need to know' basis.

Electronics has created many new opportunities for electricians, service engineers, technicians and installers and for this reason I have included the chapters on security systems and communication systems. Those involved in process control may find the chapter on sensors and transducers particularly interesting.

I would like to acknowledge the assistance given by the following manufacturers and organizations in the preparation of this book:

Crabtree Electrical Industries Limited
Farnell Components Ltd
Meggar Instruments Ltd (AVO)
M.K. Electric Ltd
Multicore Solders Ltd
R.S. Components Ltd
Health and Safety Executive

I would like to acknowledge my gratitude to the Open University for my own electronics education, the proposal reviewers and my colleagues at Blackpool and The Fylde College for their suggestions and assistance during the preparation of the manuscript.

Finally, I would like to thank Joyce, Samantha and Victoria for their support and encouragement.

Trevor Linsley,
Poulton-le-Fylde   January 2000

1

HEALTH AND SAFETY

Many hundreds of workers each day become the victim of an accident at work. Some accidents are simply inconvenient as well as painful, such as trapping a finger or dropping a heavy object on one's foot. Many other accidents lead to hospital treatment or time away from work, and about 100 people each year die as a result of an industrial accident. Most accidents could be avoided and are the result of workers' ignorance, neglect, forgetfulness or recklessness. It is against this background that successive governments have introduced legislation and passed laws aimed at improving safety in the workplace.

The Health and Safety at Work Act 1974

The most important piece of recent legislation has been the Health and Safety at Work Act of 1974. The purpose of the act is to provide a legal framework for stimulating and encouraging high standards of health and safety at work. The Act was the result of recommendations made by a royal commission in 1970. The commission looked at the health and safety of employees at work and concluded that the main cause of accidents was apathy on the part of both the employer and employee. The 1974 Act puts the responsibility for safety at work on both the employers and the workers.

The employer has a duty to care for the health and safety of employees (Section 2 of the act). To do this the employer must ensure that:

- the working conditions and standard of hygiene are appropriate;
- the plant, tools and equipment are properly maintained;
- the necessary safety equipment, such as personal protective equipment, dust and fume extractors and machine guards are available and properly used;
- the workers are trained to use equipment and plant safely.

Employees have a duty to care for their own health and safety and that of others who may be affected by their actions (Section 7 of the act). To do this they must:

- take reasonable care to avoid injury to themselves or others as a result of their work activity;
- co-operate with their employer, helping him or her to comply with the requirements of the act;
- not interfere with or misuse anything provided to protect their health and safety.

Failure to comply with the Health and Safety at Work Act is a criminal offence and any infringement of the law can result in heavy fines, a prison sentence or both.

ENFORCEMENT

Laws and rules must be enforced if they are to be effective. The system of control under the Health and Safety at Work Act comes from the Health and Safety Executive (HSE) which is charged with enforcing the law. The HSE is divided into a number of specialist inspectorates or sections which operate from local offices throughout the UK. From the local offices the inspectors visit individual places of work.

The HSE inspectors have been given wide-ranging powers to assist them in the enforcement of the law. They can:

- enter premises unannounced, carry out investigations and take measurements or photographs;
- take statements from individuals;
- check the records and documents required by legislation;
- give information and advice to an employee or employer about safety in the workplace;
- demand the dismantling or destruction of any equipment, materials or substance likely to cause immediate serious injury;
- issue an improvement notice which will require an employer to put right, within a specified period of time, a minor infringement of the legislation;
- issue a prohibition notice which will require an employer to stop immediately any activity likely to result in serious injury and which will be enforced until the situation is corrected;
- prosecute all persons who fail to comply with their safety duties, including employers, employees, designers, manufacturers, suppliers and the self-employed.

SAFETY DOCUMENTATION

Under the Health and Safety at Work Act, the employer is responsible for ensuring that adequate instruction and information is given to employees to make them safety-conscious. Part 1, Section 3 of the act instructs all employers to prepare a written 'Health and Safety Policy Statement' and to bring this to the notice of all employees. To promote adequate health and safety measures the employer must consult with the employees' safety representatives. In companies which employ more than 20 people this is normally done by forming a safety committee which is made up of a safety officer and employee representatives, usually nominated by a trade union. The safety officer is usually employed full-time in that role. Small companies might employ a safety supervisor, who will have other duties within the company; or alternatively they could join a 'safety group'. The safety group then shares the cost of employing a safety adviser or safety officer, who visits each company in rotation. An employee who identifies a dangerous situation should initially report to his or her site safety representative. The safety representative should then bring the dangerous situation to the notice of the safety committee for action which will remove the danger. This may mean changing company policy or procedures or making modifications to equipment. All actions of the safety committee should be documented and recorded as evidence that the company takes seriously its health and safety policy. Even small organisations employing five or more people must display a 'Health and Safety Law Poster – What You Should Know.' Following a recent revision by the Health and Safety Executive (HSE) the current poster reference number ISBN 0717613801 must be replaced by a new updated version ISBN 0717624935. All workplaces must display the new poster from 30th June 2000.

Under the general protective umbrella of the Health and Safety at Work Act, other pieces of legislation also affect people at work.

The Control of Substances Hazardous to Health Regulations 1988 (COSHH)

The regulations control people's exposure to hazardous substances in the workplace. Regulation 6 requires employers to assess the risks to health from working with hazardous substances, to train employees in techniques which will reduce the risk or provide personal protective equipment so that employees will not endanger themselves or others through exposure to hazardous substances.

Employers should also know what cleaning, storage and disposal procedures are required and what emergency procedures to follow. All this information must be available to anyone using hazardous substances and the documentation made available to a visiting HSE inspector.

Hazardous substances include:

- any substance which gives off fumes causing headaches or respiratory irritation (e.g. cleaning fluids);
- man-made fibres which might cause skin or eye irritation (e.g. fibre-glass matting);
- acids causing skin burns and breathing irritation (e.g. car batteries, which contain dilute sulphuric acid);

- solvents causing skin and respiratory irritation (e.g. strong solvents such as those used to cement together plastic component parts);
- fumes and gases causing asphyxiation (e.g. burning plastic and PVC give off toxic fumes);
- flour, cement and wood dust causing breathing problems and eye irritation.

When using hazardous substances:

- follow the company's work instruction and if you are unsure ask your safety supervisor;
- read hazard warning signs and labels – they will tell you if a substance is poisonous, easy to set on fire or causes burns to the skin;
- before you use a substance, find out what to do if it spills on to your clothes or skin or working surface;
- do not transfer small quantities of any liquids or substances into unlabelled or wrongly labelled containers.

Where personal protective equipment is provided by an employer, employees have a duty to use it to safeguard themselves.

Personal Protective Equipment (PPE)

Under the Health and Safety at Work Act, employers must provide free of charge any personal protective equipment and employees must make full and proper use of it. Safety signs such as those shown at Fig. 1.1 are useful reminders of the type of PPE to be used in a particular area. The vulnerable parts of the body which may need protection are the head, eyes, ears, lungs, torso, hands and feet and, additionally, protection from falls may need to be considered. Objects falling from a height present the major hazard against which head protection is provided. Other hazards include striking the head against projections and hair becoming entangled in machinery. Typical methods of protection include helmets, light duty scalp protectors called 'bump caps' and hairnets.

The eyes are very vulnerable to liquid splashes, flying particles and light emissions such as ultraviolet light, electric arcs and lasers. Types of eye protectors

Figure 1.1 Safety signs showing type of PPE to be worn.

include safety spectacles, safety goggles and face shields. Screen based workstations are being used increasingly in industrial and commercial locations by all types of personnel. Working with VDUs (visual display units) can cause eye strain and fatigue and, therefore, this hazard is the subject of a separate section later in this chapter headed VDU operation hazards.

Noise is accepted as a problem in most industries and surprisingly there has been very little control legislation. The Health and Safety Executive have published a 'Code of Practice' and 'Guidance Notes' HSG 56 for reducing the exposure of employed persons to noise. A continuous exposure limit of below 90 dB for an eight hour working day is recommended by the code.

Noise may be defined as any disagreeable or undesirable sound or sounds, generally of a random nature, which do not have clearly defined frequencies. The usual basis for measuring noise or sound level is the decibel scale. Whether noise of a particular level is harmful or not also depends upon the length of exposure to it. This is the basis of the widely accepted limit of 90 dB of continuous exposure to noise for eight hours per day.

A peak sound pressure of above 200 pascals or about 120 dB is considered unacceptable and 130 dB is the threshold of pain for humans. If a person has to shout to be understood at two metres, the background noise is about 85 dB. If the distance is only one metre, the noise level is about 90 dB. Continuous noise at work causes deafness, makes people irritable, affects concentration, causes fatigue and accident proneness and may mask sounds which need to be heard in order to work efficiently and safely.

It may be possible to engineer out some of the noise, for example by placing a generator in a separate sound-proofed building. Alternatively, it may be possible to provide job rotation, to rearrange work locations or provide acoustic refuges.

Where individuals must be subjected to some noise at work it may be reduced by ear protectors. These may be disposable ear plugs, re-usable ear plugs or ear muffs. The chosen ear protector must be suited to the user and suitable for the type of noise and individual personnel should be trained in its correct use.

Breathing reasonably clean air is the right of every individual, particularly at work. Some industrial processes produce dust which may present a potentially serious hazard. The lung disease asbestosis is caused by the inhalation of asbestos dust or particles and the coal dust disease pneumoconiosis, suffered by many coal miners, has made people aware of the dangers of breathing in contaminated air.

Some people may prove to be allergic to quite innocent products such as flour dust in the food industry or wood dust in the construction industry. The main effect of inhaling dust is a measurable impairment of lung function. This can be avoided by wearing an appropriate mask, respirator or breathing apparatus as recommended by the company's health and safety policy and indicated by local safety signs such as those shown in Fig. 1.2.

A worker's body may need protection against heat or cold, bad weather, chemical or metal splash, impact or penetration and contaminated dust. Alternatively, there may be a risk of the worker's own clothes causing contamination of the product, as in the food industry. Appropriate clothing will be recommended in the company's health and safety policy and may include conventional or disposable overalls, boilersuits, warehouse coats, donkey jackets, aprons or some other specialist clothing. Figure 1.3 shows typical PPE safety signs to be found in the food industry.

Hands and feet may need protection from abrasion, temperature extremes, cuts and punctures, impact or skin infection. Gloves or gauntlets provide protection from most industrial processes but should not be worn when operating machinery because they may become entangled in it. Care in selecting the appropriate protective device is required; for example, barrier creams provide only a limited protection against infection.

Boots or shoes with in-built toe caps can give protection against impact or falling objects and, when fitted with a mild steel sole plate, can also provide protection from sharp objects penetrating through the

Figure 1.2 Breathing protection signs.

Figure 1.3 PPE and safety signs to be found in the food industry.

sole. Special slip resistant soles can also be provided for employees working in wet areas.

Whatever the hazard to health and safety at work, the employer must be able to demonstrate that he or she has carried out a risk analysis, made recommendations which will reduce that risk and communicated these recommendations to the workforce. Where there is a need for PPE to protect against personal injury and to create a safe working environment, the employer must provide that equipment and any necessary training which might be required and the employee must make full and proper use of such equipment and training.

The Electricity at Work Regulations 1989 (EWR)

This legislation came into force in 1990 and replaced earlier regulations such as the Electricity (Factories Act) Special Regulations 1944. The purpose of the regulations is to 'require precautions to be taken against the risk of death or personal injury from electricity in work activities.'

Section 4 of the EWR tells us that:

> all systems must be constructed so as to prevent danger ... , and be properly maintained. ... Every work activity shall be carried out in a manner which does not give rise to danger. ... In the case of work of an electrical nature, it is preferable that the conductors be made dead before work commences.

The EWR do not tell us specifically how to carry out our work activities and ensure compliance but if proceedings were brought against an individual for breaking the EWR, the only acceptable defence would be 'to prove that all reasonable steps were taken and all diligence exercised to avoid the offence' (Regulation 29). An electronics service engineer could reasonably be expected to have 'exercised all diligence' if the installation was wired according to the IEE Wiring Regulations (see below).

The IEE Wiring Regulations

The Institution of Electrical Engineers Requirements for Electrical Installations (the IEE Regulations) are non-statutory regulations. They relate principally to the design, selection, erection, inspection and testing of electrical installations, whether permanent or temporary, in and about buildings generally and to agricultural and horticultural premises, construction sites and caravans and their sites. Paragraph 7 of the introduction to the EWR says:

> The IEE Wiring Regulations is a code of practice which is widely recognised and accepted in the United Kingdom and compliance with them is likely to achieve compliance with all relevant aspects of the Electricity at Work Regulations. The IEE Wiring Regulations only apply to installations operating at a voltage up to 1000 V a.c. They do not apply to electrical installations in mines and quarries, where special regulations apply because of the adverse conditions experienced there.

The current edition of the IEE Wiring Regulations is the 16th, incorporating amendment numbers 1 (1994) and 2 (1997). The main reason for incorporating the IEE Wiring Regulations into British Standard BS 7671 was to create harmonisation with European standards. Electronics service engineers can reasonably be expected by their customers and employers to carry out their work to the safe standards detailed in the IEE Regulations.

British and European Standards marks

Goods manufactured to the exacting specifications laid down by the British Standards Institution (BSI) are suitable for the purpose for which they were made. There seems to be a British Standard for practically everything made today and compliance with the relevant British Standard is, in most cases, voluntary. However, when specifying or installing equipment, the electrical designer or service engineer needs to be sure that the materials are suitable for their purpose and offer a degree of safety and should only use equipment which carries the appropriate British Standards number.

The British Standards Institution has created two important marks of safety; the BSI kite mark and the BSI safety mark, which are shown in Fig. 1.4. The BSI kite mark is an assurance that the product carrying the label has been produced under a system of supervision, control and testing and can only be used by manufacturers who have been granted a licence under the scheme. It does not necessarily cover safety unless the appropriate British Standard specifies a safety requirement. The BSI safety mark is a guarantee of the product's electrical, mechanical and thermal safety. It does not guarantee a product's performance.

The CE mark is not a quality mark but an indication given by the manufacturer or importer that the product or system meets the legal safety requirements of the European Commission and can, therefore, be presumed safe to use. The mark is applied by the manufacturer after carrying out the appropriate tests to ensure compliance with the relevant safety standards. The CE mark shown in Fig. 1.5 gives the manufacturer the right to sell the product in all the countries of the European Community. All electrical products installed after January 1997 must bear the CE mark.

Figure 1.5 European Commission safety mark.

Electrical safety in workshops and laboratories

Electric shock is generally caused by either touching a conductor that is normally live, called direct contact, or by touching a metal part made live by an electrical fault, called indirect contact. Electrical supplies at voltages above 50 V a.c. can kill human beings and livestock and must, therefore, be treated with caution and respect. The touch voltage curve of Fig. 1.6 shows that a person in contact with the domestic mains voltage of 230 V must be released from this dangerous situation in 40 ms if harmful effects are to be avoided. (See also the section on Electric Shock later in this chapter.) Similarly, a person in contact with an industrial three phase supply of 400 V must be released in 15 ms to avoid being harmed.

BSI kite mark

BSI safety mark

Figure 1.4 BSI kite and safety marks.

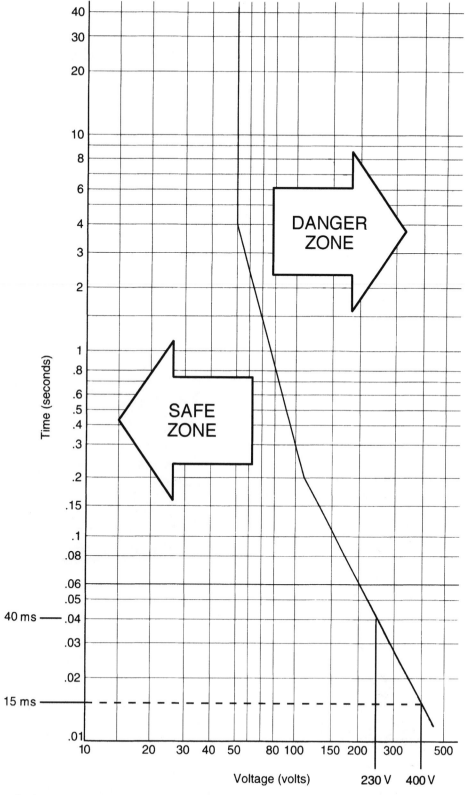

Figure 1.6 Touch voltage curve.

In general, protection against direct contact with live parts is achieved by insulating those parts. Protection against indirect contact is achieved by earthing and bonding and the automatic disconnection of the supply in the event of a fault occurring by fuses, miniature circuit breakers (MCBs) or residual circuit devices (RCDs).

OVERCURRENT PROTECTION

The electrical mains equipment to the laboratory or workshop must provide protection against overcurrent (IEE Regulation 431). Fuses provide overcurrent protection when connected in the live conductor; they *must not* be connected in the neutral conductor. Miniature circuit breakers may be used in place of fuses and these have the added advantage of being able to reset a faulty circuit at the flick of a switch. A fuse and MCB are shown in Fig. 1.7.

Overcurrent can be sub-divided into overload current and short circuit current. An overload current can be defined as a current which exceeds the rated value in an otherwise healthy circuit. Overload current usually occurs because the circuit is being abused or 'overloaded' and may result in currents of two or three times the rated current flowing in the circuit. A short circuit is an overcurrent resulting from a fault between live and neutral conductors or between live conductors and earth. Short circuit currents usually occur as a result of an accident which could not have been predicted before the event. Short circuit currents may be hundreds of times greater than the rated current. In both overload and short circuit currents the basic requirements for protection are that the fault currents should be quickly interrupted and the particular circuit isolated safely before the fault current causes a rise in temperature which might damage the insulation and terminations of the electrical installation. Fuses and MCBs provide this protection.

RESIDUAL CURRENT DEVICE (RCD)

An RCD is an electrical safety device which constantly monitors the balance of the current through two coils connected to the live and neutral conductors of the load. In a healthy circuit these live and neutral currents balance, but if a fault occurs the balance is lost and a trip circuit opens a double pole switch to isolate the load. Modern RCDs designed to

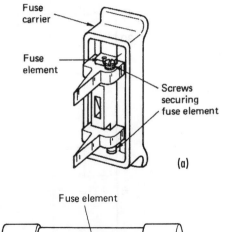

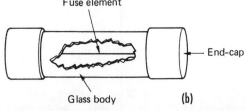

Figure 1.7 Overcurrent protection devices (a) a semi-enclosed fuse (b) a cartridge fuse and (c) an MCB.

protect people have tripping current sensitivities of 30 mA and, therefore, a faulty circuit can be isolated very quickly before the lethal limit to human beings of 50 mA is reached. (See the section on Electric Shock later in this chapter.)

Electrical supply distribution fuseboards can be supplied which incorporate an RCD, and this type of protection is now normally provided for all workshop and laboratory installations. However, where this type of protection is not provided, for example in domestic installations, remote garages or when using portable tools or equipment away from the electrical service workshop, a 'plug-in' RCD such as that shown in Fig. 1.8 can provide the high levels of human protection described above.

Figure 1.8 A plug-in RCD for safe electrical assembly.

Another source of danger in electronic assembly and repair is the soldering iron which may cause burns or start a fire. This danger can be reduced by always using a soldering iron stand or an appropriate type of soldering iron. These are further discussed in Chapter 2.

In general terms, a workshop or laboratory used for the testing and servicing of electrical equipment must have an adequate number of socket outlets usually protected by an RCD where portable equipment is to be repaired and tested.

Secure isolation

As a service engineer working on electrical equipment you must always make sure that the equipment is switched off or electrically isolated before commencing work. Every circuit must be provided with a means of isolation (IEE Regulation 130–06–01). When working on portable equipment or desk top units it is often simply a matter of unplugging the equipment from the adjacent supply. Larger pieces of equipment, such as main servers or uninterruptible power supplies (UPS), may require isolating at the local isolator switch before work commences. To deter anyone from re-connecting the supply while work is being carried out on equipment, a sign 'Danger – Service Engineer at Work' should be displayed on the isolator and the isolation 'secured' with a small padlock or the fuses removed so that no-one can re-connect whilst work is being carried out on that piece of equipment. The Electricity at Work Regulations 1989 are very specific at Regulation 12(1) that we must ensure the disconnection and separation of electrical equipment from every source of supply and that this disconnection and separation is secure. Where a test instrument or voltage indicator is used to prove the supply dead, Regulation 4(3) of the Electricity at Work Regulations 1989 recommends that the following procedure is adopted.

1 First connect the test device such as that shown in Fig. 1.9 to the supply which is to be isolated. The test device should indicate mains voltage.
2 Next, isolate the supply and observe that the test device now reads zero volts.

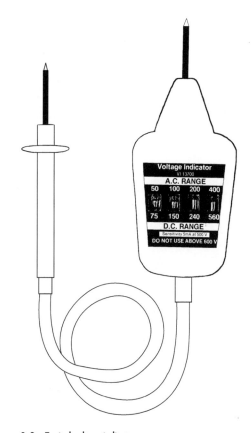

Figure 1.9 Typical voltage indicator.

3 Then connect the same test device to a known live supply or proving unit such as that shown in Fig. 1.10 to 'prove' that the tester is still working correctly.
4 Finally secure the isolation and place warning signs; only then should work commence.

The test device being used by the service engineer must incorporate safe test leads which comply with the Health and Safety Executive Guidance Note 38 on electrical test equipment. These leads should incorporate barriers to prevent the service engineer touching live terminals when testing a protective fuse and be well insulated and robust, such as those shown in Fig. 1.11.

To isolate a piece of equipment or individual circuit successfully, competently, safely and in accordance with all the relevant regulations, we must follow a procedure such as that given by the flow diagram of Fig. 1.12. Start at the top and work down the flow diagram. When the heavy outlined boxes are reached, pause and ask yourself whether everything is satisfactory up to this point. If the answer is 'yes', move on. If the answer is 'no', go back as indicated by the diagram.

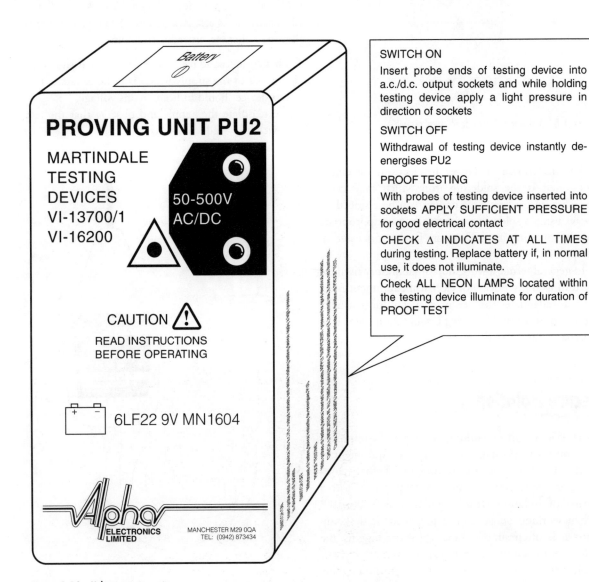

Figure 1.10 Voltage proving unit.

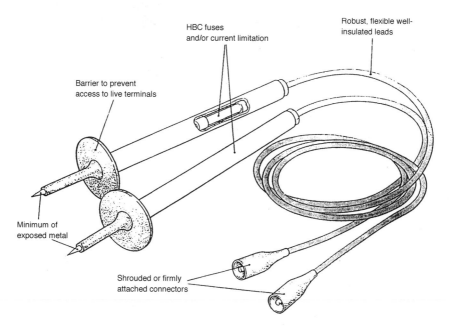

Figure 1.11 Recommended type of test probe and leads.

LIVE TESTING

The Electricity at Work Regulations 1989 at Regulation 4(3) tell us that it is preferable that supplies be made dead before work commences. However, it does acknowledge that some work, such as fault finding and testing, may require the electrical equipment to remain energised. Therefore, if the fault finding and testing can only be successfully carried out live then the person carrying out the fault diagnosis must:

- be trained so that they understand the equipment and the potential hazards of working live and can, therefore, be deemed 'competent' to carry out that activity;
- only use approved test equipment;
- set up appropriate warning notices and barriers so that the work activity does not create a situation dangerous to others.

While live testing may be required by electronics service engineers in order to find the fault, live repair work must not be carried out. The individual circuit or piece of equipment must first be isolated before work commences in order to comply with the Electricity at Work Regulations 1989.

Risk and hazard

The Health and Safety at Work Act refers extensively to 'risk', 'hazard' and 'competent persons'. They have specific meanings in relation to the law and we should, therefore, consider them.

- A *hazard* is anything that can cause harm, for example chemicals, electricity, working at heights, etc.
- *Risk* is the chance, big or small, of harm actually being done as a result of the potential hazard.
- *Competent persons* are often referred to in the Health and Safety at Work Regulations, but who is 'competent'? For the purposes of the act, a competent person is anyone who has the necessary technical skills, training and expertise to safely carry out the particular activity. Therefore, a competent person dealing with a *hazardous* situation reduces the *risk*.

The rules and regulations of the safe working environment are communicated to employees by written instructions, signs and symbols and by other employees as they go about their daily work. All signs are intended to inform and give warning of possible

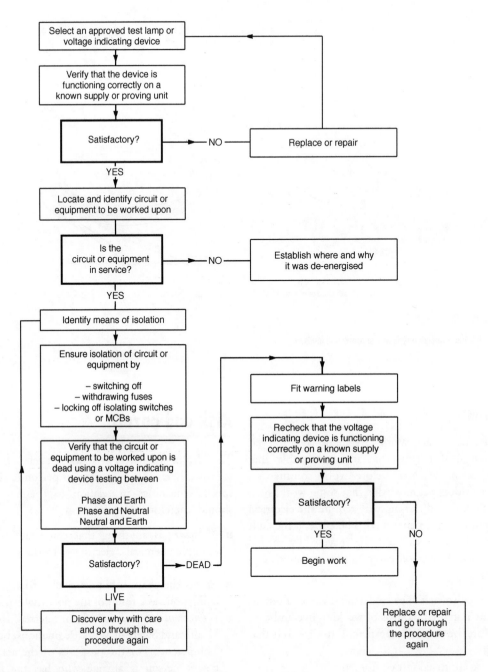

Figure 1.12 Flowchart for a secure isolation procedure.

dangers and should be obeyed. Warning signs such as those shown in Fig. 1.13 warn of hazardous situations.

In the food manufacturing industry, process equipment is stripped down, washed and cleaned at the end of each product run. This results in a large amount of surface water on the floor around the machinery during cleaning operations. This is a potential slipping *hazard*. If the area is screened off during the cleaning operation to prevent unauthorised access, the *risk* of someone being harmed is reduced.

Alternatively, think of a service engineer opening a network file server to carry out a minor repair. Electricity is the *hazard* and there is a potential *risk* of electric shock to the service engineer and the

Figure 1.13 Warning signs give warning of a hazardous situation.

'crashing' of the whole system if a screwdriver accidentally 'shorts' out one of the live conductors. Isolating the server as described previously or carrying out the repair 'out of hours' when the electricity supply could be made 'dead' would greatly reduce the risks associated with this simple repair.

HAZARD RISK ASSESSMENT

In the working environment things change and move on. New materials and equipment come in and personnel move to other jobs. When you walk around your workplace think about what might go wrong at each stage of what you do. For example, here are some typical activities and the possible hazards associated with them.

- delivery of materials – lifting and carrying
- stacking and storage – falling materials
- movement of personnel and materials – tripping, falling, collisions
- processing raw materials – exposure to toxic substances
- maintenance of buildings – working at heights, falling
- maintenance of equipment – lifting and moving
- working with electricity – electric shock
- movement of vehicles – collisions and breakdowns
- dealing with emergencies – spillage, breakages, fire
- noise and lighting levels – injuries to employees

In industrial process plants, most accidents are caused by a few key activities. In most commercial, service and light industrial companies the hazards are few and relatively trivial. However, let us concentrate on those hazards which could cause serious harm.

In order to make a decision about risks, think about the worst possible result; is it likely to be a broken finger, permanent lung damage or death? Then think about how often an accident is likely to occur as a result of a particular hazard. How often is the job producing this particular hazard done? How close do personnel get to the hazard? How likely is it that something can go wrong? Finally, consider how many people might get hurt if things do go wrong. Might this also include people who do not work for the company, visitors, contractors or the general public?

The Health and Safety Executive recommend five steps to risk assessment in their publication 'A Step by Step Guide to a Safer and Healthier Workplace.'

Step 1 recommends that you look for the hazards in your workplace

Step 2 recommends that you then decide who might be harmed by that hazard and how harm might occur

Step 3 asks you to evaluate the risk arising from the hazard and to decide whether existing precautions are adequate or whether more should be done. Employers must first ask themselves if they have done everything that the law requires to be done. Then they should ask if they are using generally accepted industrial safety standards and finally, if they are doing everything that is 'reasonably practicable to keep the workplace safe'? The real aim is to make all risks as small as possible. If there is a risk, ask yourself:

- can I get rid of the hazard altogether?
- if not, how can I control the risk so that harm is unlikely to occur?

Personal protection equipment should only be used to control a risk when there is nothing else that can reasonably be done.

Step 4 tells us to record the findings. If a company employs fewer than five employees they are not required to write anything down but companies who employ five or more people must record any significant findings of the risk assessment and identify any further action required to control the risk. A simple procedure based upon these four steps is all that is required.

Step 5 tell us to review the assessment from time to time and revise the risk assessment if necessary. This is because, as previously mentioned, things change in the working environment; new equipment, machines and procedures can lead to new hazards.

Hazard and risk assessment will be discussed later in this chapter in relation to VDU operating hazards. HSE publications may be obtained from HSE Books, P.O. Box 1999, Sudbury, Suffolk, CO10 6FS.

Safety signs

The rules and regulations of the working environment are communicated to employees by written instructions, signs and symbols. All signs in the working environment are intended to inform. They should give warning of possible dangers and must be obeyed. At first there were many different safety signs but British Standard BS 5378 Part 1 (1980) and the Health and Safety (Signs and Signals) Regulations 1996 have introduced a standard system which gives health and safety information with the minimum use of words. The purpose of the regulations is to establish an internationally understood system of safety signs and colours which draw attention to equipment and situations that do, or could, affect health and safety. Text-only safety signs became illegal from 24th December 1998. From that date, all safety signs have had to contain a pictogram or symbol such as those shown in Fig. 1.14.

Safety signs fall into four categories; mandatory, prohibition, warning or safe condition.

MANDATORY SAFETY SIGNS

These are circular blue signs with a white symbol or pictogram as shown in Figs. 1.1 and 1.2. They give *instructions* which must be obeyed.

PROHIBITION SAFETY SIGNS

These are circular white signs with a red border and red cross bar with a black symbol or pictogram as shown in Fig. 1.15. They indicate an activity which *must not be done*.

WARNING SAFETY SIGNS

These are triangular yellow signs with a black border and symbol or pictogram as shown in Fig. 1.13. They give *warning* of a hazard or danger.

Figure 1.14 Text only safety signs do not comply.

Figure 1.15 Prohibition signs.

SAFE CONDITION SAFETY SIGNS

These are square or rectangular green signs with a white symbol or pictogram as shown in Fig. 1.16. They give *information* about the safety provision.

First aid at work

Despite all the safety precautions taken at work accidents do happen and you may be the only other person able to take action to assist a workmate. If you are not a qualified first aider limit your help to obvious common sense assistance and call for help but do remember that if a workmate's heart or breathing has stopped as a result of an accident he or she has only minutes to live unless you act quickly.

THE LEGAL REQUIREMENTS

The Health and Safety (First Aid) Regulations 1981 and relevant approved codes of practice and guidance notes place a duty of care on all employers to provide *adequate* first aid facilities appropriate to the type of work being undertaken. Adequate facilities will relate to a number of factors such as:

- how many employees are employed?
- what type of work is being carried out?
- are there any special or unusual hazards?
- are employees working in scattered and/or isolated locations?
- is there shift work or 'out of hours' work being undertaken?
- is the workplace remote from emergency medical services?
- are there inexperienced workers on site?
- what were the risks of injury and ill health identified by the company's Hazard Risk Assessment?

The regulations state that:

Employers are under a duty to provide such numbers of suitable persons as is *adequate and appropriate in the circumstances* for rendering first aid to his employees if they are injured or become ill at work. For this purpose a person shall not be suitable unless he or she has undergone such training and has such qualifications as the Health and Safety Executive may approve.

This is typical of the way in which the health and safety regulations are written. The regulations and codes of practice do not specify numbers, but set out guidelines in respect of the number of first aiders needed, dependent upon the type of company, the hazards present and the number of people employed.

FIRST AID

Let us now consider the questions 'what is first aid?' and 'who might become a first aider?' The regulations give the following definitions of first aid. '*First aid* is the treatment of minor injuries which would otherwise receive no treatment or do not need treatment by a doctor or nurse' *or* 'In cases where a person will require help from a doctor or nurse, first aid is treatment for the purpose of preserving life and minimising the consequences of an injury or illness until such help is obtained.' A more generally accepted definition of first aid might be as follows: *first aid* is the initial assistance or treatment given to a casualty for any injury or sudden illness before the arrival of an ambulance, doctor or other medically qualified person.

Now having defined first aid, who might become a first aider? A *first aider* is someone who has undergone a training course to administer first aid at work and holds a current first aid certificate. The training

 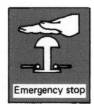 

Figure 1.16 Safe condition signs.

course and certification must be approved by the HSE. The aims of a first aider are to preserve life, to limit the worsening of the injury or illness and to promote recovery.

A first aider may also undertake the duties of an *appointed person*. An *appointed person* is someone who is nominated to take charge when someone is injured or becomes ill, including calling an ambulance if required. The appointed person will also look after the first aid equipment, including re-stocking the first aid box.

Appointed persons should not attempt to give first aid for which they have not been trained but should limit their help to obvious common sense assistance and summon professional assistance as required. Suggested numbers of first aid personnel are given in Table 1.1. The actual number of first aid personnel must take into account any special circumstances such as remoteness from medical services, the use of several separate buildings and the company's hazard risk assessment. First aid personnel must be available at all times when people are at work, taking into account shift working patterns and providing cover for sickness absences.

Every company must have at least one first aid kit under the regulations. The size and contents of the kit will depend upon the nature of the risks involved in the particular working environment and the number of employees. Table 1.2 gives a list of the contents of any first aid box to comply with the HSE Regulations.

There now follows a description of some first aid procedures which should be practised under expert guidance before they are carried out in an emergency.

Bleeding

External bleeding is easily recognised and treatment is relatively simple. Apply pressure to the site of the wound using the fingers or a dressing and keep that pressure on for 10 minutes. If possible, raise the injury above the level of the victim's heart and make him or her rest. Tourniquets are no longer advocated to control bleeding. To avoid possible contact with hepatitis or the AIDS virus when dealing with open wounds, first aiders should avoid contact with fresh blood by either wearing plastic gloves or by allowing the casualty to apply pressure to the bleeding wound.

Internal bleeding cannot be treated by first aiders but it must be recognised so that the casualty can be rushed to hospital. The symptoms are that the casualty looks very ill, with pallid, cold and clammy skin; breathing becomes fast and shallow; the casualty becomes restless and anxious and the heart beat, which can be monitored by taking the pulse, becomes faster and weaker.

Burns

The general principle for the treatment of burns is the removal of heat from the affected area to relieve pain. Therefore, if possible, relieve the pain by placing the injured part under clean cold water for about 10 minutes.

Do not remove burnt clothing sticking to the skin and *do not* apply lotions or ointments. *Do not* break blisters or attempt to remove loose skin. Cover the injured area with a clean sterile dressing. Burn victims with a serious injury should be escorted to hospital

Table 1.1 Suggested numbers of first aid personnel.

Category of risk	Numbers employed at any location	Suggested number of first aid personnel
Lower risk e.g. shops and offices, libraries	Fewer than 50 50–100 More than 100	At least one appointed person At least one first aider One additional first aider for every 100 employed
Medium risk e.g. light engineering and assembly work, food processing, warehousing	Fewer than 20 20–100 More than 100	At least one appointed person At least one first aider for every 50 employed (or part thereof) One additional first aider for every 100 employed
Higher risk e.g. most construction, slaughterhouses, chemical manufacture, extensive work with dangerous machinery or sharp instruments	Fewer than five 5–50 More than 50	At least one appointed person At least one first aider One additional first aider for every 50 employed

Table 1.2 Contents of first aid boxes.

Item	No. of employees				
	1–5	6–10	11–50	51–100	101–150
Guidance card on general first aid	1	1	1	1	1
Individually wrapped sterile adhesive dressings	10	20	40	40	40
Sterile eye pads, with attachment (Standard Dressing No. 16 BPC)	1	2	4	6	8
Triangular bandages	1	2	4	6	8
Sterile covering for serious wounds (where applicable)	1	2	4	6	8
Safety pins	6	6	12	12	12
Medium sized sterile unmedicated dressings (Standard Dressings No. 9 and No. 14 and the Ambulance Dressing No. 1)	3	6	8	10	12
Large sterile unmedicated dressings (Standard Dressings No. 9 and No. 14 and the Ambulance Dressing No. 1)	1	2	4	6	10
Extra large sterile unmedicated dressings (Ambulance Dressing No. 3)	1	2	4	6	8

Where tap water is not available, sterile water or sterile normal saline in disposable containers (each holding a minimum of 300 ml) must be kept near the first aid box. The following minimum quantities should be kept:

Number of employees.

1–10	11–50	51–100	101–150
1 × 300 ml	3 × 300 ml	6 × 300 ml	6 × 300 ml

because the shock associated with burns may not become apparent for an hour or two after the injury and the victim may not realise how ill he or she is.

Foreign bodies in the eye

If the object cannot be removed readily with a clean piece of moist material, irrigate the eye with clean, cool water. People with eye injuries which are more than minimal must be sent to hospital with the eye covered with an eye pad.

Broken bones

Make the casualty as comfortable as possible. In the case of a broken limb, support should be given to the limb by hand or with padding. Do not move the casualty unless by remaining in that position he or she is likely to suffer further injury. Obtain professional help as soon as possible.

Choking

The traditional thump on the back is still the first aid treatment of choice for choking. In the case of children, hold them upside down and thump the back. If that does not work, sweep the back of the throat with a finger to remove the obstruction.

If all else fails, try the Heimlich Manoeuvre, which is carried out by putting the arms round the casualty from behind and placing the fist above the navel and below the rib cage. The idea is to force out the obstruction by pressure of air from the casualty's lungs. The fist is jabbed upwards towards the diaphragm, creating the pressure.

Serious injuries

Unless you are a qualified first aider, limit your help to the obvious common sense assistance and *shout loudly for help*.

If the casualty has suffered cardiac arrest or stopped breathing, immediately call the emergency services by dialling 999 and ask for the ambulance service. Give precise directions to the scene of the accident. The casualty stands the best chance of survival and recovery if the emergency services can get a rapid response paramedic team quickly to the scene. They have extensive training and will have specialist equipment with them. *This is the latest thinking (2000) on emergency procedures.* Only then should you apply mouth to mouth resuscitation or CPR (cardio-pulmonary resuscitation) until professional help arrives or the patient recovers.

The top priority is to make sure that oxygen is reaching the brain, otherwise a casualty will quickly die. Oxygen is breathed into the lungs and transferred to the blood for circulation to every organ of the body. Any interference with that process threatens life.

Airway

When people suddenly become unconscious, the tendency is to fall on to their backs. This can cut off their air supply because the tongue, lacking muscle control, falls to the back of the throat. The airway is restored by extending the head back and supporting the chin. This action lifts the tongue away from the throat. Other obstructions have to be removed by hand.

Breathing stopped

Just clearing the airway may allow breathing to restart, but if this does not happen the first aider needs to start mouth to mouth resuscitation, inflating the casualty's lungs at 15 breaths per minute. Loosen any tight clothing around the neck, chest and waist. To ensure a good airway, lay the casualty on his back and support the shoulders on some padding, tilt the head backwards and open the mouth. If the casualty is breathing faintly, lifting the tongue clear of the airway may be all that is necessary to restore normal breathing. However, if the casualty does not begin to breathe, close the casualty's nose by pinching with your fingers, open your mouth wide and take a deep breath. Seal your lips around his mouth and blow into his lungs until the chest rises. Remove your mouth and watch the casualty's chest fall. Continue this procedure at your natural breathing rate. If the mouth is damaged or you are having difficulty making a seal around the casualty's mouth, close the mouth and inflate the lungs through his nostrils. Give artificial respiration until natural breathing is restored or until professional help arrives.

Cardiac arrest – no blood circulation

The first aider needs to be able to take a pulse at both the wrist and the neck (carotid). This needs practice because initially it is not always easy to find even a strong pulse of a healthy volunteer. A complete lack of pulse is a sign of cardiac arrest and a signal for the first aider to start the rather dramatic treatment of external chest compression or CPR (cardio-pulmonary resuscitation). CPR involves pressing the breast bone down 35–50 mm so that the heart is squeezed and the circulation artificially maintained. This is coupled with mouth to mouth breathing to keep a flow of oxygen to the brain and other organs.

Other obvious signs of cardiac arrest are that the casualty's lips may be blue, the pupils of the eye widely dilated and the neck pulse cannot be felt. In these circumstances act quickly and lay the casualty on his or her back. Kneel down beside them and place the heel of one hand in the centre of the chest. Cover this hand with your other hand and interlace the fingers. Straighten your arms and press down on the chest sharply with the heel of your hands and then release the pressure. Continue to do this 15 times at the rate of one push per second. Check the casualty's pulse; if none is felt, give two breaths of artificial respiration and then a further 15 chest compressions. Continue this procedure until the heartbeat is restored, and carry on the artificial respiration until normal breathing returns. Pay close attention to the condition of the casualty whilst giving CPR. When a pulse is restored the blueness around the mouth will quickly go away and you should stop the CPR. Look carefully at the rate of breathing. When this is also normal, stop giving artificial respiration. Treat the casualty for shock and obtain professional help. The skills of cardio-pulmonary resuscitation need to be practised on a mannequin dummy before a real life emergency presents itself, because the position of the hands on the patient's chest is critical.

Electric shock

Cardiac arrest sometimes happens following a severe electric shock and the symptoms should be treated with CPR as described above. To a healthy person an electric shock from the mains is equivalent to a severe blow on the chest. It may make you jump back very suddenly and leave you breathless. Switch off the supply and sit quietly while trying to discover why you received the shock. To the very young or very old, or someone with a less robust constitution, an electric shock from the mains can be serious. When this happens, it is necessary to act quickly to prevent the electric shock becoming fatal.

Upon finding someone receiving an electric shock, do not touch the person whilst they are still in contact with the electrical supply or you will risk being electrocuted yourself.

1. Switch off the supply or pull out the plug or remove the person from the supply without touching them, e.g. push or pull them off with a broom, dry towel or coat.
2. If breathing or heartbeat has stopped, immediately call professional help or instruct someone to dial 999 and then apply mouth to mouth resuscitation and CPR until the patient recovers or professional help arrives.
3. Treat for shock.
4. Transport to hospital for medical assistance if required.

Shock

Everyone suffers from shock following an accident. The severity of the shock depends upon the nature and extent of the injury. In cases of severe shock the casualty will become pale and his skin clammy from sweating. He may feel faint, have blurred vision, feel sick and complain of thirst. Reassure the casualty that everything that needs to be done is being done. Loosen tight clothing and keep him or her warm and dry until help arrives. *Do not* move the casualty unnecessarily or give anything to drink.

Accident reports

Every accident must be reported to an employer and the details of the accident and treatment given suitably documented. A first aid log book such as that shown in Fig. 1.17 containing first aid treatment record sheets could be used to effectively document accidents which occur in the workplace and the treatment given. Failure to do so may influence the payment of compensation at a later date if an injury leads to permanent disability.

RIDDOR

RIDDOR stands for Reporting of Injuries, Diseases and Dangerous Occurrences Regulation 1995, which

Figure 1.17 First Aid logbook.

is sometimes referred to as RIDDOR 95, or just RIDDOR for short. The HSE requires employers to report some work related accidents or diseases so that they can identify where and how risks arise, investigate serious accidents and publish statistics and data to help reduce accidents at work.

What needs reporting? Every work related death, major injury, dangerous occurrence, disease or any injury which results in an absence from work of over three days.

Where an employee or member of the public is killed as a result of an accident at work the employer or his representative must report the accident to the Environmental Health Department of the local authority by telephone that day and give brief details. Within 10 days this must be followed up by a complete accident report form (Form No. F2508). Major injuries sustained as a result of an accident at work include amputations, loss of sight (temporary or permanent), fractures to the body other than to fingers, thumbs or toes and any other serious injury. Once again, the Environmental Health Department of the local authority must be notified by telephone on the day that the serious injury occurs and the telephone call followed up by a completed Form F2508 within 10 days. Dangerous occurrences are listed in the regulations and include the collapse of a lift, an explosion or injury caused by an explosion, the collapse of a scaffold over five metres high, the collision of a train with any vehicle, the unintended collapse of a building and the failure of fairground equipment.

Depending upon the seriousness of the event, it may be necessary to immediately report the incident to the local authority. However, the incident must be reported within 10 days by completing Form F2508. If a doctor notifies an employer that an employee is suffering from a work related disease then form F2508A must be completed and sent to the local authority. Reportable diseases include certain poisonings, skin diseases, lung disease, infections and occupational cancer. The full list is given within the pad of report forms.

An accident at work resulting in an over three day injury, that is, an employee being absent from work for over three days as a result of an accident at work, requires that accident report form F2508 be sent to the local authority within 10 days.

An over three day injury is one which is not major but results in the injured person being away from work for more than three days not including the day the injury occurred.

Who are the reports sent to? They are sent to the Environmental Health Department of the local authority or the area HSE offices. Accident report forms F2508 can also be obtained from them or by ringing the HSE Infoline. The relevant telephone numbers and addresses are given in Appendix O at the back of this book.

For most businesses, a reportable accident, dangerous occurrence or disease is a very rare event. However, if a report is made, the company must keep a record of the occurrence for three years after the date on which the incident happened. The easiest way to do this would probably be to file a photocopy of the completed accident report form F2508, but a record may be kept in any form which is convenient.

Fire control

A fire is a chemical reaction which will continue if fuel, oxygen and heat are present. To eliminate a fire *one* of these components must be removed. This is often expressed as the fire triangle shown in Fig. 1.18 where all three corners of the triangle must be present for a fire to burn.

FUEL

Fuel is found in the work environment in many forms. Petrol and paraffin for portable generators and

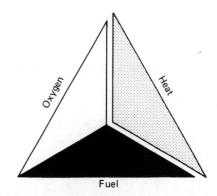

Figure 1.18 The fire triangle.

heaters, bottled gas for heating and soldering and most solvents are flammable. Rubbish also represents a source of fuel; off-cuts of wood, rags, empty solvent cans and discarded packaging will all provide fuel for a fire.

To eliminate fuel as a source of fire, all flammable liquids and gases should be stored correctly, usually in an outside locked store. The working environment should be kept clean by placing rags in a metal bin with a lid. Combustible waste materials should be removed from the work site or burned outside under controlled conditions by a competent person.

OXYGEN

Oxygen is all around us in the air we breathe but can be eliminated from a small fire by smothering with a fire blanket, sand or foam. Closing doors and windows, but not locking them, will limit the amount of oxygen available to a fire in a building and help to prevent it spreading. Most substances will burn if they are given a high enough temperature and a supply of oxygen. The minimum temperature at which a substance will burn is called the 'minimum ignition temperature', and for most materials this is considerably higher than the surrounding temperature. However, a danger does exist from portable heaters, blow torches and hot air guns which provide heat and can cause a fire by raising the temperature of materials placed in their path above the minimum ignition temperature. A safe distance must be maintained between heat sources such as soldering irons and all flammable materials.

HEAT

Heat can be removed from fire by dousing with water, but water must not be used on burning liquids since the water will spread the liquid and the fire. Some fire extinguishers have a cooling action which removes heat from the fire.

Fires in industry damage property and materials, injure people and sometimes cause loss of life. Everyone should make an effort to prevent fires but those which do break out should be extinguished as quickly as possible. In the event of fire you should:

- raise the alarm;
- turn off machinery, gas and electricity supplies in the area of the fire;
- close doors and windows but do not lock or bolt them;
- remove combustible materials and fuels away from the path of the fire only if the fire is small and this can be done safely;
- attack small fires with the correct extinguisher.

Only attack the fire if you can do so without endangering your own safety in any way. Those not involved in fighting the fire should walk to a safe area or assembly point.

Fires are divided into four classes or categories:

- Class A are wood, paper and textile fires;
- Class B are liquid fires such as paint, petrol and oil;
- Class C are fires involving gas or spilled liquefied gas;
- Class D are very specialised fires involving burning metal.

Electrical fires do not have a special category because once started, they can be identified as one of the other category types.

Fire extinguishers are for dealing with small fires and different types of fire must be attacked with a different type of extinguisher. Using the wrong type of extinguisher could make matters worse. For example, water must not be used on a liquid or electrical fire. The normal procedure when dealing

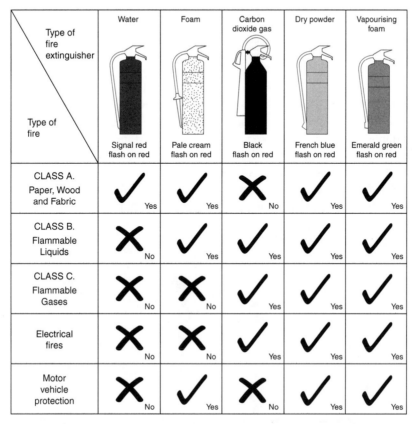

Figure 1.19 Fire extinguishers and their applications (colour codes to BS EN3: 1996). The base colour of all fire extinguishers is red, with a different coloured flash to indicate the type.

with electrical fires is to cut off the electrical supply and use an extinguisher which is appropriate to whatever is burning. Figure 1.19 shows the correct type of extinguisher to be used on five different types of fire. The colour coding of the portable fire extinguishers shown complies with BS 5423: 1980 and 1987 and BS EN3: 1996.

Manual handling

Manual handling is lifting, transporting or supporting loads by hand or by bodily force. The load might be any heavy object, a printer, a visual display unit, a box of tools or a stepladder. Whatever the heavy object is, it must be moved thoughtfully and carefully, using appropriate lifting techniques if personal pain and injury are to be avoided. Many people hurt their back, arms and feet, and over one third of all three day reported injuries submitted to the HSE each year are the result of manual handling.

When lifting heavy loads, correct lifting procedures must be adopted to avoid back injuries. Figure 1.20 demonstrates the technique. Do not lift objects from the floor with the back bent and the legs straight as this causes excessive stress on the spine. Always lift with the back straight and the legs bent so that the powerful leg muscles do the lifting work. Bend at the hips and knees to get down to the level of the object being lifted, positioning the body as close to the object as possible. Grasp the object firmly and, keeping the back straight and the head erect, use the leg muscles to raise in a smooth movement. Carry the load close to the body. When putting the object down, keep the back straight and bend at the hips and knees, reversing the lifting procedure. There have been too many injuries over the years resulting from bad manual handling techniques. The problem has become so serious that the Health and Safety Executive has introduced new legislation under the Health and Safety at Work Act 1974, the Manual Handling Operations Regulations 1992. Publications such as *Getting to Grips with Manual Handling* can be obtained from HSE Books; the address and Infoline are given in Appendix O.

Where a job involves considerable manual handling, employers must now train employees in the correct lifting procedures and provide the appropriate equipment necessary to promote the safe manual handling of loads.

Consider some 'good practice' when lifting loads.

- Do not lift the load manually if it is more appropriate to use a mechanical aid. Only lift or carry what you can easily manage.
- Always use a trolley, wheelbarrow or truck such as those shown in Fig. 1.21 when these are available.
- Plan ahead to avoid unnecessary or repeated movement of loads.
- Take account of the centre of gravity of the load when lifting – the weight acts through the centre of gravity.
- Never leave a suspended load unsupervised.
- Always lift and lower loads gently.
- Clear obstacles out of the lifting area.
- Use the manual lifting techniques described above and avoid sudden or jerky movements.
- Use gloves when manual handling to avoid injury from rough or sharp edges.
- Take special care when moving loads wrapped in grease or bubble-wrap.
- Never move a load over other people or walk under a suspended load.

Figure 1.20 Correct manual lifting and carrying procedure.

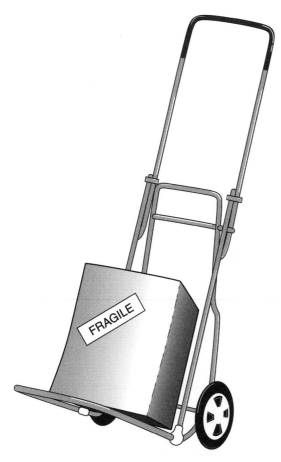

Figure 1.21 Always use a mechanical aid to transport a load when available.

Tidiness

Slips, trips and falls are still the major cause of accidents at work. To help prevent them:

- keep work areas clean and tidy;
- keep walkways clear;
- do not leave objects lying around blocking up walkways;
- clean up spills or wet patches on the floor straight away;
- fit appropriate temporary danger signs – 'danger wet floor' or 'danger slippery surface', etc. such as those shown in Fig. 1.22.

Personal hygiene

In the work environment, dirt and contact with chemicals and cleaning fluids may make you feel ill or cause unpleasant skin complaints. Therefore, you should always:

- wear appropriate personal protective equipment;
- wash your hands after using the toilet, after work and before you eat a meal, using soap and water or appropriate cleaners;
- dry your hands with the towel or dryer provided; do not use rags or your clothes;
- use barrier creams when they are provided to protect your skin;
- obtain medical advice about any skin complaint such as rashes, blisters or ulcers and tell your supervisor of the problems being experienced.

Ergonomics

Ergonomics is the scientific study of the efficiency of workers at work and the conditions required for them to achieve their maximum efficiency. To be efficient at work, workers must follow the workplace rules as described above. However, they must equally be *provided* with the tools and equipment which will enable them to be efficient in an environment which is safe and clean. For example:

- Hygiene and welfare facilities must be in place and appropriate for the workforce, such as toilets and washbasins, soap and towels, lockers, drinking water and adequate facilities for taking food and refreshment.
- Premises, furniture and fittings must be kept clean and spillage must be cleaned up.
- Floors and gangways must be clear and safe to use.
- Equipment being used by workers must be designed for good health, such as seats and machine controls designed for best control and posture.
- The workplace must be a safe place to work or, where particular hazards exist, they must be clearly identified.

Figure 1.22 For safe working – fit appropriate temporary signs.

- The workplace must be a comfortable place to work, with an adequate temperature, usually above 16°C or 60°F for sedentary occupations and good ventilation. Where workers must work in high or low temperatures (e.g. freezer rooms) adequate clothing (PPE) must be provided.
- The workplace must be adequately illuminated. This means good general illumination with no glare (see VDU operation hazards later), no flicker from fluorescent tubes and adequate emergency lighting.

Personal awareness

Every year thousands of people have accidents at their place of work despite the legal requirements laid down by the Health and Safety Executive. Many people recover quickly but an accident at work can result in permanent harm or even death.

At the very least, injuries hurt individuals. They may prevent you from doing the things you enjoy in your spare time and they cost a lot of money, to you in loss of earnings and to your employer in loss of production and possibly damage to equipment. Your place of work may look harmless but it can be dangerous.

If there are five or more people employed by your company then the company must have its own safety policy as described earlier in this chapter. This must spell out the organisation and arrangements which have been put in place to ensure that you and your workmates are working in a safe place.

Your employer must also have carried out an assessment on the risks to your health and safety in the place where you are working. You should be told about the safety policy and risk assessment, for example you may have been given a relevant leaflet when you started work.

You have a responsibility under the Health and Safety at Work Act to:

- learn how to work safely and to follow company procedures of work;
- obey all safety rules, notices and signs;
- not interfere with or misuse anything provided for safety;
- report anything that seems damaged, faulty or dangerous;
- behave sensibly, not play practical jokes and not distract other people at work;
- walk sensibly and not run around the workplace;
- use the prescribed walkways;
- drive only those vehicles for which you have been properly trained and passed the necessary test;
- not wear jewellery which could become caught in moving parts if you are using machinery at work;
- always wear appropriate clothing and PPE if necessary.

VDU operation hazards

Those who work at supermarket checkouts, assemble equipment or components or work for long periods at a visual display unit (VDU) and keyboard can be at risk because of the repetitive nature of the work. The hazard associated with these activities is a medical condition called *upper limb disorders*. The term covers a number of related medical conditions including tenosynovitis, carpal tunnel syndrome and tennis elbow which affect the arms, forearms and hands.

The symptoms of upper limb disorders include pain or soreness and limited movement of the affected parts. Typical causes are incorrect posture, fatigue brought on by too great a workload, over-forceful movements and inadequate rest periods.

Injuries from upper limb disorders can be prevented by improving the design and layout of the work areas. For example, the position of the keyboard and VDU screen, the height of the workbench and chair, the provision of simple tools, the inclusion of rest periods into the work cycle and better training and supervision will all help.

HEALTH AND SAFETY (DISPLAY SCREEN EQUIPMENT) REGULATIONS 1992

To encourage employers to protect the health of their workers and reduce the risks associated with VDU work, the Health and Safety Executive (HSE) have introduced the Health and Safety (Display Screen Equipment) Regulations 1992. The regulations came into force on 1st January 1993, and employers who use standard office VDUs must show that they have taken steps to comply with the regulations.

So who is affected by the regulations? The regulations identify employees who use VDU equipment as 'users' if they:

- use a VDU more or less continuously on most days;
- use a VDU more or less continuously for periods of an hour or more each day;
- need to transfer information quickly to or from the screen;
- need to apply high levels of attention or concentration to information displayed on a screen;
- are very dependent upon VDUs or have little choice about using them.

All VDU users must be trained to use the equipment safely and protect themselves from upper limb disorders, temporary eyestrain, headaches, fatigue and stress.

To comply with the regulations an employer must:

- train users of VDU equipment and those who will carry out a risk assessment;
- carry out a workstation risk assessment;
- plan changes of activities or breaks for users;
- provide eye and eyesight testing for users;
- make sure new workstations comply with the regulations in the future;
- give users information on the above.

User training

Good user training will normally cover the following topics:

- the operating hazards and risks as describe above;
- the importance of good posture and changing position as shown in Fig. 1.23;
- how to adjust furniture to avoid risks;
- how to organise the workstation to avoid awkward or repeated stretching movements;
- how to avoid reflections and glare on the monitor screen;
- how to adjust and clean the monitor screen;
- how to organise working routines so that there is a change of activity or a break;
- how a user might contribute to a workstation risk assessment;
- who to contact if problems arise.

Figure 1.23 Examples of good posture when using VDU equipment.

When carrying out user training, the trainer might want to consider using a video, a computer based training programme, discussions or seminars or the HSE employee leaflet *Working with VDUs* which can be obtained from the address given in Appendix O.

Workstation risk assessment

A simple way to carry out a workstation risk assessment is to use a checklist such as that shown later in this section. Users can work through the checklist themselves. They know what the problems at their workstation are and whether they are comfortable or not. A trainer/assessor should then check the completed checklist and resolve the problems which the user cannot solve. For example, users may not know how the adjustment mechanism actually operates on their chair – a shorter user may benefit from a footrest as shown in Fig. 1.23, or document holder may be more convenient for word processing users as shown in Fig. 1.24.

Breaks

Breaking up long spells of display screen work helps to prevent fatigue and upper limb problems. Where possible encourage VDU users to carry out other tasks such as taking telephone calls, filing and photocopying. Otherwise, plan for users to take breaks away from the VDU screen if possible. The length of break required is not fixed by the law; the time will vary depending upon the work being done. Breaks should be taken before users become tired and short frequent breaks are better than longer infrequent ones.

Eye and eyesight testing

VDU users and those who are to become users of VDU equipment can request an eye and eyesight test that is free of charge to them. If the test shows that they need glasses specifically to carry out their VDU work, then their employer must pay for a basic pair of frames and lenses. Users are also entitled to further tests at regular intervals but if the user's normal glasses are suitable for VDU work, then the employer is not required to pay for them.

Workstations

Make sure that new workstations comply with the regulations when:

- major changes to the workstation display screen equipment, furniture or software are made;
- new users start work or change workstations;
- workstations are re-sited;
- the nature of the work changes considerably.

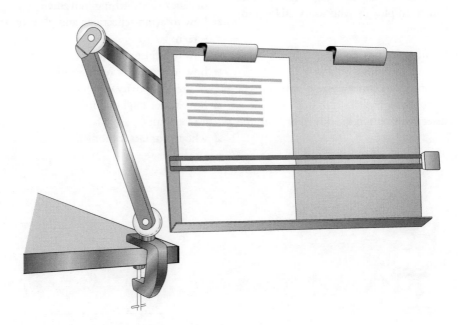

Figure 1.24 A document holder typically used by a word processing VDU operator.

Users, trainers and assessors should focus on those aspects which have changed. For example:

- if the location of the workstation has changed, is the lighting adequate, is lighting or sunlight now reflecting off the display unit?
- different users have different needs – replacing a tall user with a short user may mean that a footrest is required;
- users working from a number of source documents will need more desk space than users who are word processing.

A risk assessment should always be carried out on a new workstation or when a new operator takes over a workstation. Some questions cannot be answered until a user has had an opportunity to try the workstation. For example, does the user find the layout comfortable to operate, are there reflections on the screen at different times of the day as the sun moves around the building?

To be comfortable the operator should adjust the chair and equipment so that:

- Arms are horizontal and eyes are roughly at the height of the top of the VDU casing.
- Hands can rest on the work surface in front of the keyboard with fingers outstretched over the keys.
- Feet are placed flat on the floor – too much pressure on the backs of legs and knees may mean that a footrest is needed.
- The small of the back is supported by the chair. The back should be held straight with the shoulders relaxed.

The arms on the chair or obstructions under the desk must not prevent the user from getting close enough to the keyboard comfortably.

Information

Good employers, who comply with the Display Screen Equipment Regulations, should let their employees know what care has been taken to reduce the risk to their health and safety at work. Users should be given information on:

- the health and safety relating to their particular workstations;
- the risk assessments carried out and the steps taken to reduce risks;
- the recommended break times and changes in activity to reduce risks;
- the company procedures for obtaining eye and eyesight tests.

This information might be communicated to workers by:

- telling staff, for example, as part of an induction programme;
- circulating a booklet or leaflet to relevant staff;
- putting the information on a noticeboard;
- using a computer based information system, providing staff are trained in their use.

VDU WORKSTATION RISK ASSESSMENT CHECKLIST

Using a checklist such as that shown below or the more extensive checklist shown in the HSE book 'VDUs, An Easy Guide to the Regulations' is one way to assess workstation risks. You don't have to, but many employers find it a convenient method.

Risk factors are grouped under five headings and to each question the user should initially give a simple yes/no response. A 'yes' response means that no further action is necessary but a 'no' response will indicate that further follow-up action is required to reduce or eliminate risks to a user.

1. *Is the display screen image clear?*
1.1 Are the characters readable? Y/N
1.2 Is the image free of flicker and movement? Y/N
1.3 Are brightness and contrast adjustable? Y/N
1.4 Does the screen swivel and tilt? Y/N
1.5 Is the screen free from glare and reflections? Y/N

2. *Is the keyboard comfortable?*
2.1 Is the keyboard tiltable? Y/N
2.2 Can you find a comfortable keyboard position? Y/N
2.3 Is there enough space to rest your hands in front of the keyboard? Y/N
2.4 Are the characters on the keys easily readable? Y/N

3. *Does the furniture fit the work and user?*
3.1 Is the work surface large enough? Y/N
3.2 Is the surface free of reflections? Y/N
3.3 Is the chair stable? Y/N
3.4 Do the adjustment mechanisms work? Y/N
3.5 Are you comfortable? Y/N

4. *Is the surrounding environment risk free?*
4.1 Is there enough room to change position and vary movement? Y/N
4.2 Are levels of light, heat and noise comfortable? Y/N
4.3 Does the air feel comfortable in terms of temperature and humidity? Y/N

5. *Is the software user friendly?*
5.1 Can you comfortably use the software? Y/N
5.2 Is the software suitable for the work task? Y/N
5.3 Have you had enough training? Y/N

A copy of all risk assessments carried out should be placed in a dedicated file which can then be held by the trainer/assessor or other responsible person.

A copy of the full checklist can be found in the publication 'VDUs, an Easy Guide to the Regulations'. Other relevant publications include 'Display Screen Equipment Work and Guidance on Regulations L26' and 'Industry Advisory (General) leaflet IND(G) 36(L) 1993 Working with VDUs'. These and other Health and Safety Publications are available from the HSE; the address is given in Appendix O.

Exercises

1. The most important far-reaching recent piece of safety legislation has been:
   (a) the IEE Regulations
   (b) the Electricity at Work Regulations 1989
   (c) the Health and Safety at Work Act 1974
   (d) The British and European Standards Mark.

2. An overload current may be defined as:
   (a) a current in excess of at least 15 A
   (b) a current which exceeds the rated value in an otherwise healthy circuit
   (c) an overcurrent resulting from a fault between live and neutral conductors
   (d) a current in excess of 60 A.

3. A short circuit may be defined as:
   (a) a current in excess of at least 15 A
   (b) a current which exceeds the rated value in an otherwise healthy circuit
   (c) an overcurrent resulting from a fault between live and neutral conductors
   (d) a current in excess of 60 A.

4. RCDs or residual current devices, are designed to protect people from an electric shock and, therefore, have tripping sensitivities of about:
   (a) 10 mA
   (b) 30 mA
   (c) 50 A
   (d) 100 A.

5. RCD protection is important where:
   (a) fixed appliances are used such as VDUs
   (b) portable appliances such as soldering irons are being used
   (c) fixed appliances are used such as printers
   (d) fixed appliances are used such as desk top lamps.

6. When isolating a piece of mains equipment before carrying out a maintenance or repair procedure, the service engineer must:
   (a) follow a recognised procedure
   (b) unplug the device and get on with it
   (c) switch off the device and get on with it
   (d) work live on double insulated equipment.

7. The Electricity at Work and all other regulations tell the service engineer that he/she must:
   (a) work on electrical equipment in a 'live' condition
   (b) never work on electrical equipment in a 'live' condition
   (c) find electrical faults quickly
   (d) find electrical faults quickly but bill for the first statutory hours.

8. Hazard may be defined as:
   (a) anything that can cause harm
   (b) the chance, large or small, of harm actually being done
   (c) someone who has the necessary training and expertise to safely carry out an activity
   (d) the rules and regulations of the working environment.

9. Risk may be defined as:
   (a) anything that can cause harm
   (b) the chance, large or small, of harm actually being done
   (c) someone who has the necessary training and expertise to safely carry out an activity
   (d) the rules and regulations of the working environment.

10. A competent person may be defined as:
    (a) anything that can cause harm

(b) the chance, large or small, of harm actually being done
(c) someone who has the necessary training and expertise to safely carry out an activity
(d) the rules and regulations of the working environment.

11 First aid may be defined as:
(a) the treatment of cut fingers or toes
(b) the treatment of bruised fingers or toes
(c) the treatment of minor bleeding or abrasions
(d) the initial assistance give for any injury or illness.

12 A first aider may be defined as:
(a) an all round good chap
(b) someone who holds a current First Aid Certificate, having completed a training course
(c) someone who is nominated to take charge when someone is injured or becomes ill
(d) someone who is medically qualified.

13 An appointed person may be defined as:
(a) an all round good chap
(b) someone who holds a current First Aid Certificate, having completed a training course
(c) someone who is nominated to take charge when someone is injured or becomes ill
(d) someone who is medically qualified.

14 A small blow-torch burn to the arm of a workmate should be treated by:
(a) immersing in cold water before applying a clean dry dressing
(b) pricking any blisters before applying a clean dry dressing
(c) covering burned skin with cream or petroleum jelly to exclude the air before applying a clean dry dressing
(d) applying direct pressure to the burned skin to remove the heat from the burn and relieve the pain.

15 For any fire to continue to burn, three components must be present. These are:
(a) fuel, wood, cardboard
(b) petrol, oxygen, bottled gas
(c) flames, fuel, heat
(d) fuel, oxygen, heat.

16 A CO gas fire extinguisher, colour coded black, is suitable on:
(a) class A fires only
(b) class A and B fires only
(c) class B, C and electrical fires only
(d) electrical and motor vehicle fires only.

17 Describe the action to be taken upon finding a workmate apparently dead on the floor and connected to a live electrical supply.

18 Briefly identify the main difference between the Electricity at Work Regulations 1989 and the IEE Regulations 16th Edition (BS 7671).

19 State the responsibilities under the Health and Safety at Work Act of:
(a) an employer to his or her employees
(b) an employee to his or her employer and fellow workers.

20 Safety signs are used in the working environment to give information and warnings. Describe the purpose of the four categories of signs and state their colour code and shape. You may use sketches to illustrate your answer.

21 Describe the hazards of using VDU equipment and the methods which may be employed to reduce the risk.

22 Describe, using good manual handling techniques, how to lift a heavy object from the floor to the workbench.

23 Describe the symptoms of cardiac arrest.

24 Describe how to carry out CPR (cardio-pulmonary resuscitation).

25 Describe how to carry out mouth-to-mouth resuscitation.

2

ELECTRONIC COMPONENT RECOGNITION

There are numerous electronic components, diodes, transistors, thyristors and integrated circuits each with its own limitations, characteristics, and designed application. When repairing electronic circuits it is important to replace damaged components with an identical or equivalent component. Manufacturers issue comprehensive catalogues with details of working voltage, current, power dissipation etc., and the reference numbers of equivalent components, and some of this information is included in the Appendices. These catalogues of information, together with a high impedance multi-meter as described in Chapter 6, should form a part of the extended tool kit for a service engineer proposing to repair electronic circuits.

Electronic circuit symbols

The European Standard EN60617, which incorporates the British Standard BS 3939, recommends that particular graphical symbols be used to represent a range of electronic components on circuit diagrams. Figure 2.1 shows a selection of electronic symbols.

RESISTORS

All materials have some resistance to the flow of an electric current but, in general, the term *resistor* describes a conductor specially chosen for its resistive properties.

Resistors are the most commonly used electronic component and they are made in a variety of ways to suit the particular type of application. They are usually manufactured as either carbon composition or carbon film. In both cases the base resistive material is carbon and the general appearance is of a small cylinder with leads protruding from each end, as shown in Fig. 2.2(a).

If subjected to overload, carbon resistors usually decrease in resistance since carbon has a negative temperature coefficient. This causes more current to flow through the resistor, the temperature rises and failure occurs, usually by fracturing. Carbon resistors have a power rating of between 0.1 W and 2 W which should not be exceeded.

When larger power rated resistors are required a wire wound resistor should be chosen. This consists of a resistance wire of known value wound on a small ceramic cylinder which is encapsulated in a vitreous enamel coating as shown in Fig. 2.2(b). Wire wound resistors are designed to run hot and have a power rating up to 20 W. Care should be taken when mounting wire wound resistors to prevent the high operating temperature affecting any surrounding components.

A variable resistor is one which can be varied continuously from a very low value to the full rated resistance. This characteristic is required in tuning circuits to adjust the signal or voltage level for brightness, volume or tone. The most common type used in electronic work has a circular carbon track contacted by a metal wiper arm. The wiper arm can be adjusted by means of an adjusting shaft (rotary type) or by placing a screwdriver in a slot (preset type) as shown in Fig. 2.3. Variable resistors are also known as potentiometers because they can be used to adjust the potential difference (voltage) in a circuit. The variation in resistance can be either to a logarithmic or linear scale.

The value of the resistor and the tolerance may be

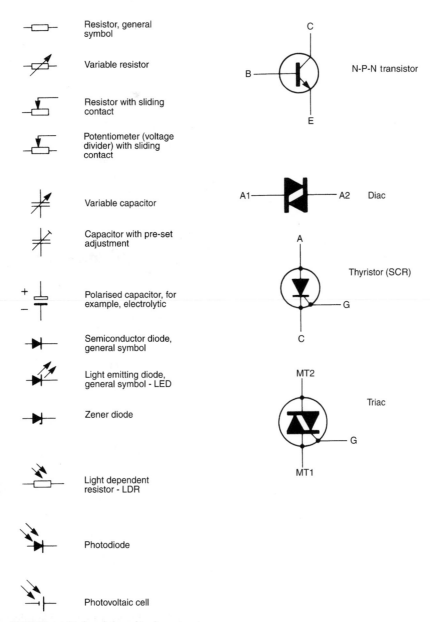

Figure 2.1 Some EN 60617 graphical symbols used in electronics.

marked on the body of the component either by direct numerical indication or by using a standard colour code. The method used will depend upon the type, physical size and manufacturer's preference, but in general the larger components have values marked directly on the body and the smaller components use the standard resistor colour code.

Abbreviations used in electronics

Where the numerical value of a component includes a decimal point, it is standard practice to include the prefix for the multiplication factor in place of the decimal point, to avoid accidental marks being

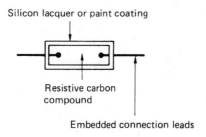

(a) Carbon composition resistor

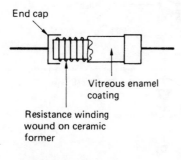

(b) Wire wound resistor

Figure 2.2 Construction of resistors.

mistaken for decimal points. Multiplication factors and prefixes are dealt with in Chapter 8.

The abbreviation R means × 1
k means × 1000
M means × 1 000 000

Therefore, a 4.7 kΩ resistor would be abbreviated to 4k7. A 5.6 Ω resistor to 5R6 and a 6.8 MΩ resistor to 6M8.

Tolerances may be indicated by adding a letter at the end of the printed code.

The abbreviation F means ± 1%
G means ± 2%
J means ± 5%
K means ± 10%
M means ± 20%

Therefore a 4.7 kΩ resistor with a tolerance of 2% would be abbreviated to 4k7G. A 5.6 Ω resistor with a tolerance of 5% would be abbreviated to 5R6J. A 6.8 MΩ resistor with a 10% tolerance would be abbreviated to 6M8K.

This is the British Standard BS 1852 code which is recommended for indicating the values of resistors on circuit diagrams and components when their physical size permits.

The standard colour code

4 BAND CARBON FILM RESISTORS

Small carbon resistors are marked with a series of four coloured bands as shown in Table 2.1. These are read according to the standard colour code to determine the resistance. The bands are located on the component towards one end. If the resistor is turned so that this end is towards the left, the bands are then read from left to right. Band (a) gives the first number of the component value, band (b) the second number, band (c) the number of zeros to be added after the first two numbers and band (d) indicates the resistor tolerance, which is commonly gold or silver, indicating a tolerance of 5% or 10% respectively. If the bands are not clearly oriented towards one end, first identify the tolerance band and turn the resistor so that this is towards the right before commencing to read the colour code as described.

The tolerance band indicates the maximum tolerance variation in the declared value of resistance. Thus a 100 Ω resistor with a 5% tolerance, will have a value somewhere between 95 Ω and 105 Ω since 5% of 100 Ω is 5 Ω.

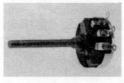

Rotary type

Preset type

Figure 2.3 Types of variable resistor.

Table 2.1 The 4 band resistor colour code

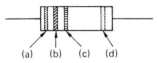

Colour	Band (a) first number	Band (b) second number	Band (c) number of noughts	Band (d) tolerance band ±
Black	0	0	None	–
Brown	1	1	1	1%
Red	2	2	2	2%
Orange	3	3	3	–
Yellow	4	4	4	–
Green	5	5	5	–
Blue	6	6	6	–
Violet	7	7	7	–
Grey	8	8	–	–
White	9	9	–	–
Gold	–	–	÷10	5%
Silver	–	–	÷100	10%
None	–	–	–	20%

EXAMPLE 1

A resistor is colour coded yellow, violet, red, gold. Determine the value of the resistor.

band (a) – yellow has a value of 4
band (b) – violet has a value of 7
band (c) – red has a value of 2
band (d) – gold indicates a tolerance of 5%

The value is therefore 4700 ± 5% and, could be written as 4.7 kΩ ± 5% or 4k7J.

EXAMPLE 2

A resistor is colour coded green, blue, brown, silver. Determine the value of the resistor.

band (a) – green has a value of 5
band (b) – blue has a value of 6
band (c) – brown has a value of 1
band (d) – silver indicates a tolerance of 10%

The value is therefore 560 ± 10% and could be written as 560 Ω ± 10% or 560RK.

EXAMPLE 3

A resistor is colour coded blue, grey, green, gold. Determine the value of the resistor.

band (a) – blue has a value of 6
band (b) – grey has a value of 8
band (c) – green has a value of 5
band (d) – gold indicates a tolerance of 5%

The value is therefore 6 800 000 ± 5% and could be written as 6.8 MΩ ± 5% or 6M8J.

EXAMPLE 4

A resistor is colour coded orange, white, silver, silver. Determine the value of the resistor.

band (a) – orange has a value of 3
band (b) – white has a value of 9
band (c) – silver indicates divide by 100 in this band
band (d) – silver indicates a tolerance of 10%

The value is therefore 0.39 ± 10% and could be written as 0.39 Ω ± 10% or R39K.

5 band metal film resistors

Metal film resistors look very like carbon film resistors but are manufactured differently and offer high precision coupled with high stability. They are often recommended for use in industrial electronic equipment. The 5 band resistor colour code is read in exactly the same way as the 4 band code but the extra band gives us a way of expressing the resistant value as a three digit number. The 5 band colour code is given in Table 2.2, but let us consider an example.

A resistor is colour coded blue, grey, black, orange and brown. Determine the value of the resistor.

band (a) – blue has a value of 6
band (b) – grey has a value of 8
band (c) – black has a value of 0
band (d) – orange has a value of 3
band (e) – brown has a value of 1

The metal film resistor has a value of 680 000 ± 1% and could, be written as 680 kΩ ± 1% or 680 kF.

Table 2.2 The 5 band resistor colour code

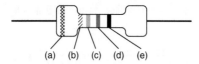

Colour	Band (a) first number	Band (b) second number	Band (c) third number	Band (d) number of noughts	Band (e) tolerance band ±
Black	0	0	0	None	–
Brown	1	1	1	1	1%
Red	2	2	2	2	2%
Orange	3	3	3	3	–
Yellow	4	4	4	4	–
Green	5	5	5	5	–
Blue	6	6	6	6	–
Violet	7	7	7	–	–
Grey	8	8	8	–	–
White	9	9	9	–	–
Gold	–	–	–	÷10	–
Silver	–	–	–	÷100	–

Table 2.3 Preferred values

E6 series 20% tolerance	E12 series 10% tolerance	E24 series 5% tolerance
10	10	10
		11
	12	12
		13
15	15	15
		16
	18	18
		20
22	22	22
		24
	27	27
		30
33	33	33
		36
	39	39
		43
47	47	47
		51
	56	56
		62
68	68	68
		75
	82	82
		91

PREFERRED VALUES

It is difficult to manufacture small electronic resistors to exact values by mass production methods. This is not a disadvantage as in most electronic circuits the value of the resistors is not critical. Manufacturers produce a limited range of *preferred* resistance values rather than an overwhelming number of individual resistance values. Therefore, in electronics, we use the preferred value closest to the actual value required.

A resistor with a preferred value of 100 Ω and a 10% tolerance could have any value between 90 Ω and 110 Ω. The next larger preferred value which would give the maximum possible range of resistance values without too much overlap would be 120 Ω. This could have any value between 108 Ω and 132 Ω. Therefore, these two preferred value resistors cover all possible resistance values between 90 Ω and 132 Ω. The next preferred value would be 150 Ω, then 180 Ω, 220 Ω and so on.

There is a series of preferred values for each tolerance level as shown in Table 2.3 so that every possible numerical value is covered. Table 2.3 indicates values between 10 and 100 but larger values can be obtained by multiplying these preferred values by a multiplication factor. Resistance values of 47 Ω, 470 Ω, 4.7 kΩ, 470 kΩ, 4.7 MΩ etc., are available in this way.

TESTING RESISTORS

The resistor being tested should have a value close to the preferred value and within the tolerance stated by the manufacturer. To measure the resistance of a resistor which is not connected into a circuit, the leads of a suitable ohm meter should be connected to each resistor connection lead and a reading obtained. The ohm meter and its use are discussed in Chapter 6.

If the resistor to be tested is connected into an electronic circuit it is *always necessary*, first to disconnect one lead from the circuit before the test leads are connected, otherwise the components in the circuit will provide parallel paths, and an incorrect reading will result.

Capacitors

The fundamental principles of capacitors and the time constant of capacitor resistor circuits are discussed in Chapter 8 under the sub-heading *Electrostatics*. In this chapter we shall consider the practical aspects associated with capacitors in electronic circuits.

A capacitor stores a small amount of electric charge; it can be thought of as a small rechargeable battery which can be quickly recharged. In electronics we are not only concerned with the amount of charge stored by the capacitor but with the way the value of the capacitor determines the performance of timers and oscillators by varying the time constant of C-R circuits.

CAPACITORS IN ACTION

If a test circuit is assembled as shown in Fig. 2.4 and the changeover switch connected to d.c. the signal lamp will only illuminate for a very short pulse as the capacitor charges. The charged capacitor then blocks any further d.c. current flow as shown by the graphs of Fig. 8.4. If the changeover switch is then connected to a.c. the lamp will illuminate at full brilliance because the capacitor will charge and discharge continuously at the supply frequency. Current is *apparently* flowing through the capacitor because electrons are moving to and fro in the wires joining the capacitor plates to the a.c. supply.

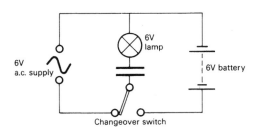

Figure 2.4 Test circuit showing capacitors in action.

Coupling and decoupling capacitors

Capacitors can be used to separate a.c. and d.c. in an electronic circuit. If the output from circuit A shown in Fig. 2.5(a) contains both a.c. and d.c. but only an a.c. input is required for circuit B then a *coupling* capacitor is connected between them. This blocks the d.c. while offering a low reactance to the a.c. component. Alternatively, if it is required that only d.c. be connected to circuit B shown in Fig. 2.5(b), a

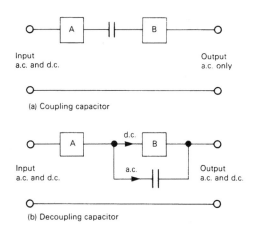

Figure 2.5 Coupling and decoupling capacitors.

decoupling capacitor can be connected in parallel with circuit B. This will provide a low reactance path for the a.c. component of the supply and only d.c. will be presented to the input of B. This technique is used to *filter out* unwanted a.c. in, for example, d.c. power supplies.

TYPES OF CAPACITOR

There are two broad categories of capacitor, the non-polarised and polarised type. The non-polarised type can be connected either way round but polarised capacitors *must* be connected to the polarity indicated otherwise a short circuit and consequent destruction of the capacitor will result. There are many different types of capacitor, each one being distinguished by the type of dielectric used in its construction. Figure 2.6 shows some of the capacitors used in electronics.

Polyester capacitors

Polyester capacitors are an example of the plastic film capacitor. Polypropylene, polycarbonate and polystyrene capacitors are other types of plastic film capacitors. The capacitor value may be marked on the plastic film or the capacitor colour code given in Table 2.4 may be used. This dielectric material gives a compact capacitor with good electrical and temperature characteristics. They are used in many electronic circuits but are not suitable for high-frequency use.

Mica capacitors

Mica capacitors have excellent stability and are accurate to ± 1% of the marked value. Since costs

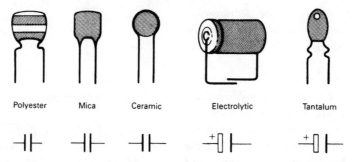

Figure 2.6 Capacitors and their symbols used in electronic circuits.

Table 2.4 Colour code for plastic film capacitors.

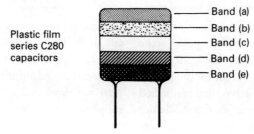

Colour	Band(a) first number	Band(b) second number	Band(c) number of noughts to be added	Band(d) tolerance	Band(e) maximum voltage
Black	–	0	None	20%	–
Brown	1	1	1	–	100 V
Red	2	2	2	–	250 V
Orange	3	3	3	–	–
Yellow	4	4	4	–	400 V
Green	5	5	5	5%	–
Blue	6	6	6	–	–
Violet	7	7	7	–	–
Grey	8	8	8	–	–
White	9	9	9	10%	–

Capacitor colour code for plastic film capacitors. Values in pF (picofarad – $\times 10^{-12}$ F)

usually increase with increased accuracy, they tend to be more expensive than plastic film capacitors. They are used where high stability is required, for example, in tuned circuits and filters.

Ceramic capacitors

Ceramic capacitors are mainly used in high-frequency circuits subjected to wide temperature variations. They have a high stability and low loss.

Electrolytic capacitors

Electrolytic capacitors are used where a large value of capacitance coupled with a small physical size is required. They are constructed on the 'Swiss roll' principle as are the paper dielectric capacitors used for p.f. correction in electrical installation circuits. The electrolytic capacitors' high capacitance for very small volume is derived from the extreme thinness of the dielectric coupled with a high dielectric strength.

Electrolytic capacitors have a size gain of approximately one hundred times over the equivalent non-electrolytic type. Their main disadvantage is that they are polarised and must be connected to the correct polarity in a circuit. Their large capacity makes them ideal as smoothing capacitors in power supplies.

Tantalum capacitors

Tantalum capacitors are a new type of electrolytic capacitor using tantalum and tantalum oxide to give a further capacitance/size advantage. They look like a 'raindrop' or 'blob' with two leads protruding from the bottom. The polarity and values may be marked on the capacitor or the colour code shown in Table 2.5 may be used. The voltage ratings available tend to be low as with all electrolytic capacitors. They are also extremely vulnerable to reverse voltages in excess of 0.3 V. This means that even when testing with an ohm meter, extreme care must be taken to ensure correct polarity.

Variable capacitors

Variable capacitors are constructed so that one set of metal plates moves relative to another set of fixed metal plates as shown in Fig. 2.7. The plates are separated by air or mica sheet which acts as the dielectric. Air dielectric variable capacitors are used to tune radio receivers to a chosen station and small variable capacitors called *trimmers* or *presets* are used to make fine, infrequent adjustments to the capacitance of a circuit.

SELECTING A CAPACITOR

When choosing a capacitor for a particular application, three factors must be considered: value, working voltage and leakage current.

The unit of capacitance is the *farad* (symbol F) to commemorate the name of the English scientist Michael Faraday. However, for practical purposes the farad is much too large and in electrical installation

Table 2.5 Colour code for tantalum polarised capacitors.

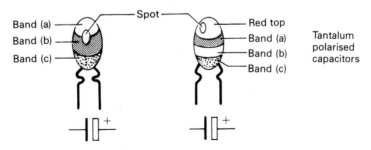

Colour	Band (a) first number	Band (b) second number	Spot number of noughts to be added	Band (c) maximum voltage
Black	—	0	None	10 V
Brown	1	1	1	—
Red	2	2	2	—
Orange	3	3	—	—
Yellow	4	4	—	6.3 V
Green	5	5	—	16 V
Blue	6	6	—	20 V
Violet	7	7	—	—
Grey	8	8	÷100	25 V
White	9	9	÷1000	30 V
Pink				35 V

Capacitor colour code for tantalum capacitors. Values in mF (microfarad × 10⁻⁶ F)

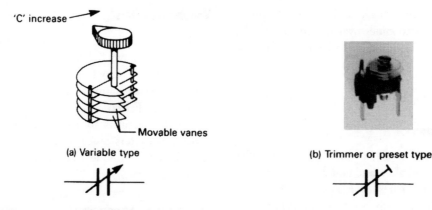

Figure 2.7 Variable capacitors and their symbols.

work and electronics we use fractions of a farad as follows:

$$1 \text{ microfarad} = 1 \text{ μF} = 1 \times 10^{-6} \text{ F}$$
$$1 \text{ nanofarad} = 1 \text{ nF} = 1 \times 10^{-9} \text{ F}$$
$$1 \text{ picofarad} = 1 \text{ pF} = 1 \times 10^{-12} \text{ F}$$

The p.f. correction capacitor used in a domestic fluorescent luminaire would typically have a value of 8 μF at a working voltage of 400 V. In an electronic filter circuit a typical capacitor value might be 100 pF at 63 V.

One microfarad is one million times greater than one picofarad. It may be useful to remember that

$$1000 \text{ pF} = 1 \text{ nF, and } 1000 \text{ nF} = 1 \text{ μF}$$

The working voltage of a capacitor is the *maximum* voltage that can be applied between the plates of the capacitor without breaking down the dielectric insulating material. This is a d.c. rating and, therefore, a capacitor with a 200 V rating must only be connected across a maximum of 200 V d.c. Since a.c. voltages are usually given as rms values, a 200 V a.c. supply would have a maximum value of about 283 V which would damage the 200 V capacitor. When connecting a capacitor to the 240 V mains supply we must choose a working voltage of about 400 V because 240 V rms is approximately 340 V maximum. The 'factor of safety' is small and, therefore, the working voltage of the capacitor must not be exceeded.

An ideal capacitor which is isolated will remain charged forever, but in practice no dielectric insulating material is perfect, and the charge will slowly *leak* between the plates, gradually discharging the capacitor. The loss of charge by leakage through it should be very small for a practical capacitor.

Capacitor colour code

The actual value of a capacitor can be identified by using the colour codes given in Tables 2.4 and 2.5 in the same way that the resistor colour code was applied to resistors.

EXAMPLE 1

A plastic film capacitor is colour coded from top to bottom, brown, black, yellow, black, red. Determine the value of the capacitor, its tolerance and working voltage.

band (a) – brown has a value 1
band (b) – black has a value 0
band (c) – yellow indicates multiply by 10 000
band (d) – black indicates 20%
band (e) – red indicates 250 V

The capacitor has a value of 100 000 pF or 0.1 μF with a tolerance of 20% and a maximum working voltage of 250 V.

EXAMPLE 2

Determine the value, tolerance and working voltage of a polyester capacitor colour coded from top to bottom, yellow, violet, yellow, white, yellow.

band (a) – yellow has a value 4,
band (b) – violet has a value 7
band (c) – yellow indicates multiply by 10 000

band (d) — white indicates 10%,
band (e) — yellow indicates 400 V

The capacitor has a value of 470 000 pF or 0.47 μF with a tolerance of 10% and a maximum working voltage of 400 V.

EXAMPLE 3

A plastic film capacitor has the following coloured bands from its top down to the connecting leads: blue, grey, orange, black, brown. Determine the value, tolerance and voltage of this capacitor.

band (a) — blue has a value 6
band (b) — grey has a value 8
band (c) — orange indicates multiply by 1000
band (d) — black indicates 20%
band (e) — brown indicates 100 V

The capacitor has a value of 68 000 pF or 68 nF with a tolerance of 20% and a maximum working voltage of 100 V.

CAPACITANCE VALUE CODES

Where the numerical value of the capacitor includes a decimal point, it is standard practice to use the prefix for the multiplication factor in place of the decimal point. This is the same practice which we used earlier for resistors.

The abbreviation μ means microfarad
n means nanofarad
p means picofarad

Therefore, a 1.8 pF capacitor would be abbreviated to 1p8, a 10pF capacitor to 10p, a 150 pF capacitor to 150p or n15, a 2200 pF capacitor to 2n2 and a 10 000 pF capacitor to 10n.

$$1000 \text{ pF} = 1 \text{ nF} = 0.001 \text{ μF}$$

TESTING CAPACITORS

The discussion earlier about *ideal* and *leaky* capacitors provides us with a basic principle to test for a faulty capacitor.

Non-polarised capacitors

Using an ohm meter as described in Chapter 6, connect the leads of the capacitor to the ohm meter and observe the reading. If the resistance is less than about 1 MΩ, it is allowing current to pass from the ohm meter and, therefore, the capacitor is leaking and is faulty. With large-value capacitors (in the μF range) there may be a short initial burst of current as the capacitor charges up.

Polarised capacitors

It is essential to connect the *true positive* of the ohm meter to the positive lead of the capacitor. When first connected, the resistance is low but rises to a steady value as the dielectric forms between the capacitor plates.

Inductors and transformers

An inductor is a coil of wire wound on a former having a core of air or iron. When a current flows through the coil a magnetic field is established. A transformer consists of two coils wound on a common magnetic core and, therefore, in this sense, the transformer is also an inductor. Simple transformer theory is discussed in Chapter 8. A small electronic transformer and the aerial of a radio receiver comprising a coil wound on a ferrite core are shown in Fig. 2.8.

Inductors such as the radio receiver aerial can be connected in parallel with a variable capacitor and *tuned* for maximum response so that a particular radio station can be listened to while excluding all others.

(a) Transformer

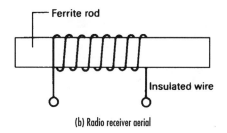

(b) Radio receiver aerial

Figure 2.8 Examples of an inductor.

Most electronic circuits require a voltage between 5 and 12 V and the transformer provides an ideal way of initially reducing the mains voltage to a value which is suitable for the particular electronic circuit.

When compared with other individual electronic components, inductors are large. The magnetic fields produced by industrial electronic equipment such as electromagnets, relays and transformers can cut across other electronic components and cause undesirable emfs to be induced. This causes electrical interference – called *electrical noise*, and may prevent the normal operation of the electronic circuit. This interference can be avoided by magnetically *screening* the inductive components from the remaining circuits.

Switches

In electrical installation work we identify four separate types of switching: switching for isolation, mechanical maintenance, emergency switching and functional switching. In electronics we are principally concerned with functional switching which is defined as the switching of electrically operated equipment in normal service. In any switch, metal contacts are brought together or separated by the action of the switch either to make or break the circuit. Figure 2.9 shows examples of some switches used in electronics.

Manufacturers' technical specifications rate switches according to their current rating, maximum working voltage and type of supply. The switch chosen for a particular application must be capable of interrupting the total steady current. This is the current rating quoted for most switches and relays and usually assumes a purely resistive load. If the load is inductive the switch must be derated by something between 25% and 50% if the safe working life of the switch is to be maintained.

The voltage rating indicates the maximum safe working voltage and is determined by the type of insulation material used in the construction of the switch and the contact separation. Again a purely resistive load is assumed, and because inductive loads may cause high voltages and current surges as the magnetic flux collapses, the switch must be derated when used with inductive loads to prevent flash-over damaging the insulation and reducing the life of the switch.

The arc established across a switch contact tends to extinguish itself every half cycle as the voltage falls to zero on an a.c. supply. A d.c. arc tends to maintain itself until contact separation becomes too great. For this reason switch ratings are usually lower when used on d.c. supplies.

At voltages below about 12 V the current rating of the switch can usually be increased without seriously reducing its working life. Switch life is usually based upon a minimum of 10 000 operations at the maximum rating for a purely resistive load. This covers both electrical and mechanical wear.

Switches have a different number of contact configurations known as poles and throws as shown in Table 2.6. The poles (P) are the number of separate circuits which the switch makes or breaks at the same time. The throws (T) are the number of positions to which each pole can be switched. A one-way lighting switch allows us to change between two stable states, on and off, and is therefore called a single-pole single-throw (SPST) switch. A two-way lighting switch moves a single pole between two contacts and is, therefore, called a single-pole, double-throw (SPDT)

(a) Toggle switch

(b) Slide switch

Figure 2.9 Switches used in electronics.

Table 2.6 Switch contact configurations

SPST		Single-pole single-throw (single-pole on-off)
SPDT		Single-pole double-throw (single-pole changeover)
DPST		Double-pole single-throw (double-pole on-off)
DPDT		Double-pole double-throw (double-pole changeover)

switch or changeover switch. A single switch toggle can be made to move two poles at the same time leading to double-pole single-throw (DPST) and double-pole, double-throw (DPDT) switches.

When using switches for the first time it is wise to use an ohm meter to trace out which contacts are joined when the switch is operated.

MICROSWITCHES

A microswitch is a small sensitive mechanical switch usually fitted with a lever or actuator so that only a small force is required to operate a snap action switch. The actuator may be a simple lever or incorporate a roller as shown in Fig. 2.10. The actuator causes contacts in the switch to open or close in various switch configurations from SPST to DPDT.

REED SWITCHES

A reed switch has two thin strips of steel (reeds) sealed inside a glass tube usually containing nitrogen or some other chemically inert gas to reduce sparking at the contacts, as shown in Fig. 2.11. The reeds are arranged so that their ends overlap but are a short distance apart. If a magnet is brought close to the reed switch the contacts attract each other and make electrical contact, closing the switch. When the magnet is removed the reeds separate and the circuit is broken. The reed switch is a *proximity* switch since it is operated by the nearness of the magnet. It can operate very quickly, 2000 times per minute, over a lifetime of more than 1000 million switching operations. Reed switches are used in telephone exchanges and door or window proximity switches in alarm systems. It also has industrial applications. Because the reeds and electrical contacts are encapsulated in a glass envelope, it can be used in explosive atmospheres.

Electromagnetic relay

An electromagnetic relay is simply an electromagnet operating a number of switch contacts as shown in Fig. 2.12. When a current is passed through the coil, the soft iron core becomes magnetised, attracts the iron armature and closes the switch contacts. The relay coil is electrically insulated from the switch contacts and, therefore, a relay is able to switch circuits operating at a different voltage to the coil operating voltage. The small current which energises the coil is also able to switch larger currents at the switch contacts. The switch part of the relay may have many poles controlling several circuits at once.

Miniature plug-in relays are popular in electronic circuits and intruder alarm circuits. However, all mechanical–electrical switches are limited in their speed of operation by the time taken physically for a movable contact to make or break a switch contact. Where extremely high-speed operations are required, the switching action must take place without physical movement. This is only possible using the properties of semiconductor materials in devices such as transistors and thyristors. They permit extremely high-speed switching without arcing and are considered in Chapter 4.

Figure 2.10 Microswitches.

Figure 2.11 Reed switch.

Figure 2.12 An electromagnetic relay.

Overcurrent protection

Every piece of electronic equipment must incorporate some means of overcurrent protection. The term overcurrent can be sub-divided into *overload* current and *short circuit* current. An overload can be defined as a current which exceeds the rated value in an otherwise healthy circuit and a short circuit as an overcurrent resulting from a fault of negligible impedance between conductors. An overload may result in currents of two or three times the rated current flowing in the circuit, while short circuit currents may be hundreds of times greater than the rated current. In both cases the basic requirement for safety is that the fault current should be interrupted quickly and the circuit isolated from the supply. Fuses provide overcurrent protection when connected in the live conductor; they must not be connected in the neutral conductor. Circuit breakers may be used in place of fuses and the best protection of all is obtained when the equipment is connected to a residual current device. Figure 2.13 shows a cartridge fuse holder. Protection from excess current is covered in some detail in Chapter 6 of *Basic Electrical Installation Work*.

Figure 2.13 A fuse holder for cartridge fuses up to 15 A.

Supplies

The a.c. mains supply is probably the cheapest and most reliable source of electrical power. Low-voltage supplies are available from signal generators and power supply units which plug into the a.c. mains and these are considered in Chapter 6. However, there are occasions when the mains is unavailable, either because it has failed as in an emergency lighting scheme, or because the electronic equipment is to be used in a remote place. In these cases we must resort to portable battery power.

BATTERIES
The zinc-carbon

The zinc-carbon battery is the cheapest available battery. It has a zinc negative electrode, a manganese dioxide positive electrode and the electrolyte is a solution of ammonium chloride. They are available in the same range of sizes as the alkaline batteries shown in Fig. 2.14, AAA, AA, C, D and PP3 having an emf of 1.5 V. This is the most popular cell for low-current use such as torches. When the chemical reaction is exhausted the battery must be replaced.

The alkaline-manganese

The alkaline-manganese battery has a zinc negative electrode, a manganese dioxide positive electrode and the electrolyte is a strong solution of the alkali potassium hydroxide. This gives the battery up to four times the energy content of the standard zinc-carbon battery. They are available in a range of sizes as shown in Fig. 2.14 and have an emf of 1.5 V. The battery is leak-proof being encased in a steel case and is ideal for use in calculators, personal radio cassettes, electric shavers and cameras. When the chemical reaction is exhausted the battery must be replaced.

The silver and mercury button cells

The silver and mercury button cells are small button-sized batteries with an emf respectively of 1.5 V and

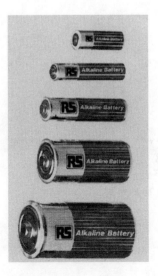

Figure 2.14 Alkaline batteries.

1.35 V. They have a long life for such a small battery and are used where small occasional currents are required, such as in watches and cameras. Silver and mercury button cells are shown in Fig. 2.15. When the chemical reaction is exhausted the battery must be replaced.

Batteries which must be replaced when the chemical reaction is exhausted are collectively called *primary cells*. A battery which can be repeatedly recharged is called a *secondary cell*, the most familiar of which is the lead-acid battery used in motor vehicles. *Lead-acid batteries* have lead plates immersed in a dilute solution of sulphuric acid. Each cell produces an emf of 2 V and most commonly six cells are grouped together in one 12 V battery. To maintain the battery in good condition the plates must always be covered and if necessary the electrolyte must be topped up with distilled water. The state of charge of a lead-acid battery can best be found by measuring the specific gravity of the electrolyte. When fully charged the specific gravity should have a value of 1.28 and when discharged or 'flat' a value of 1.15. Batteries should not be allowed to remain flat otherwise the lead sulphate hardens and cannot be changed back into lead dioxide and lead. They should be trickle charged to maintain a healthy state of readiness. Lead-acid batteries are used extensively on vehicles, as emergency supplies in public buildings and as a portable source of power in caravans. The advantage of a lead-acid battery is that it can be recharged when exhausted, but the major disadvantages are that they are very heavy and have a liquid electrolyte. The Nicad battery overcomes these difficulties and is suitable for electronic applications.

Figure 2.15 Silver and mercury button cells.

The nickel-cadmium battery (Nicad)

The nickel-cadmium battery (Nicad) has a nickel positive electrode, cadmium negative electrode and the electrolyte is potassium hydroxide. They are available in the same range of sizes as the alkaline battery and are shown in Fig. 2.16.

They have a high energy content but the terminal voltage is slightly lower than the equivalent size of alkaline battery, 1.25 V for a Nicad and 1.5 V for the equivalent alkaline. They can supply high currents, will operate successfully over a wide temperature range and are capable of accepting considerable overcharging when used with recommended chargers. Each battery will accept a minimum of 700 full discharge/charge cycles. They are used in the place of alkaline batteries where a rechargeable capability is required, for example in self-contained emergency lighting luminaires.

Figure 2.16 Nickel-cadmium rechargeable batteries.

Packaging electronic components

When we talk about packaging electronic components we are not referring to the parcel or box which contains the components for storage and delivery, but to the type of encapsulation in which the tiny semiconductor material is contained.

Figure 2.17 shows three different package outlines for just one type of discrete component, the transistor. Identification of the pin connections for different packages is given within the text as each separate or discrete component is considered, particularly in Chapter 4, Semiconductor devices. However, the Appendices aim to draw together all the information on pin connections and packages for easy reference.

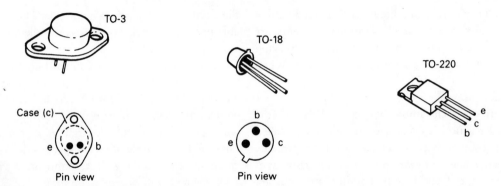

Figure 2.17 Three different package outlines for transistors.

Obtaining information and components

Electronic service engineers and electricians use electrical wholesalers and suppliers to purchase electrical cable, equipment and accessories. Similar facilities are available in most towns and cities for the purchase of electronic components and equipment. There are also a number of national suppliers who employ representatives who will call at your workshop to offer technical advice and take your order. Some of these national companies also offer a 24-hour telephone order and mail order service. Their full-coloured, fully illustrated catalogue also contains an enormous amount of technical information. The names and addresses of these national companies are given in Appendix A. For local suppliers you must consult your local telephone directory and *Yellow Pages*. The Appendices of this book also contain some technical reference information.

3
ELECTRONIC CIRCUIT ASSEMBLY

To get a 'feel' for electronics you should take the opportunity to build some of the simple circuits described in Chapter 5 using the constructional methods described in this chapter. Practical electronics can be carried out with very few tools and little resources. A kitchen table, suitably protected, or a small corner of the electrical workshop is all that is required. The place chosen should be well lit, have a flat and dry area of about 1 m × 1 m and have access to a three-pin socket.

Working with others can also be a valuable source of inspiration and encouragement. Many technical colleges and evening institutes offer basic electronic courses which give someone new to electronics an opportunity to use the tools and equipment under guidance and at little cost. The City and Guilds of London offer many electronics examinations which are particularly suitable for electricians and electronic service engineers who require a formal qualification in electronics.

Statutory regulations

In the early days of electrical services, poor design and installation led to many buildings being damaged by fire, and people and livestock being electrocuted. It was the insurance companies who originally drew up a set of rules and guidelines of good practice in the interests of reducing the number of claims made upon them. The first rules were made by the American Board of Fire Underwriters and were quickly followed by the Phoenix Rules of 1882. In the same year the first edition of the Rules and Regulations for the Prevention of Fire Risk arising from Electrical Lighting were issued by the Institution of Electrical Engineers.

The current edition of these regulations is called the Regulations for Electrical Installations and from January 1993 we have been using the 16th Edition. All the rules have been revised, updated and amended at regular intervals to take account of modern developments and the 16th Edition brings the UK Regulations into harmony with Europe under British Standard BS 7671. These now incorporate amendment numbers 1 (1994) and 2 (1997). The electrician and service engineer is now controlled by at least five sets of rules, regulations and standards. These are:

- the Electricity Supply Regulations, 1988;
- the Electricity at Work Regulations 1989;
- the IEE Regulations for Electrical Installations;
- British Standards;
- the Health and Safety at Work Act, 1974.

These rules, regulations and standards are considered in detail in Chapter 1.

Safety precautions

Electricity can be dangerous. It can give a serious shock and it does cause fires. For maximum safety, the sockets being used for the electronic test and assembly should be supplied by a residual current circuit breaker. These sense fault currents as low as 30 mA so that a faulty circuit or piece of equipment can be isolated before the lethal limit to human beings of about 50 mA is reached. Plug-in residual current devices (RCDs) of the type shown in Fig. 3.1 can now be bought very cheaply from any good

Figure 3.1 A plug-in RCD for safe electrical assembly.

electrical supplier or DIY outlet. All equipment should be earthed and fitted with a 2 A or 5 A fuse which is adequate for most electronics equipment. Larger fuses reduce the level of protection.

Another source of danger in electronic assembly is the hot soldering iron, which may cause burns or even start a fire. The soldering iron should always be placed in a soldering iron stand when not being used. The chances of causing a fire or burning yourself can be reduced by storing the soldering iron in its stand at the back of the workspace so that you don't have to lean over it when working.

So far we have been discussing the sensible safety precautions which everyone working with electricity should take. However, you or someone else in your workplace may receive an electric shock and therefore guidance is offered for dealing with such cases in Chapter 1, where electric shock is discussed in detail under the sub-heading First aid.

Hand tools

Tools extend the physical capabilities of the human body. Good-quality, sharp tools are important to any craftsman. An electrician or electronic service engineer is no less a craftsman than a wood carver. Each must work with a high degree of skill and expertise and each must have sympathy and respect for the materials which they use. The basic tools required by anyone working with electrical equipment are those used to strip, cut and connect conductors and components. The tools required for successful electronic assembly are wire strippers,

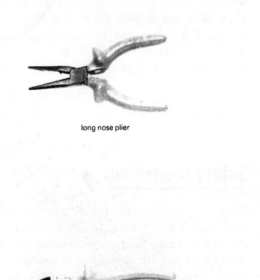

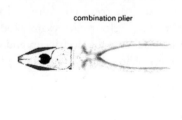

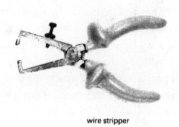

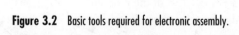

Figure 3.2 Basic tools required for electronic assembly.

diagonal cutters or snips, long nose pliers and ordinary or combinational pliers. Electricians and electronic service engineers have traditionally chosen insulated hand tools.

Figure 3.2 shows the basic hand tools required for successful electronic assembly.

Soldering irons

An electric soldering iron with the correct size bit is essential for making good-quality, permanent connections in electronic circuits. A soldering iron consists of a heat-insulated handle, supporting a heating element of between 15 W and 25 W. The bit is inserted into this element and heats up to a temperature of about 210°C by conduction. Various sizes of bits are available and are interchangeable.

Copper bits can be filed clean or rubbed with emery cloth until the tip is a bright copper colour. *Ironclad* bits must *not* be cleaned with a file or emery cloth but should be rubbed clean when they are hot, using a damp cloth or wet sponge. Before the soldering iron can be used to make electrical connections, the bit must be '*tinned*' as follows:

1. first clean the bit as described above;
2. plug in the soldering iron and allow it to heat up;
3. apply cored solder to the clean hot bit;
4. wipe off the excess solder with a damp cloth or damp sponge.

This will leave the soldering iron brightly 'tinned' and ready to be used. Figure 3.3 shows a 240 V general-purpose soldering iron and stand suitable for use in electronic assembly.

Soldering gun

Soldering guns of the type shown in Fig. 3.4 are trigger-operated soldering irons. Within ten seconds of pressing the trigger the bit is at the working temperature of 315°C. The working temperature can be arrived at even more quickly with constant use. The plastic case holds a 240 V transformer having an isolated low-voltage high-current secondary circuit which is completed by the copper soldering bit. The bits are interchangeable and should be tinned and used in the same way as the general-purpose iron considered above.

Butane gas-powered soldering irons are also available. In appearance they are very similar to the general-purpose soldering iron shown in Fig. 3.3, but without the mains cable. The handle acts as the fuel tank, various sizes of soldering bits are available and a protective cap is supplied to cover the hot end of the tool when not in use. The advantage of a gas soldering iron is that it can be used when a mains supply is not easily available.

The final choice of soldering iron will be influenced by many factors, frequency of use, where used, personal preference and cost. In 1999 the relevant costs were approximately £12 for the general-purpose iron, £25 for the soldering gun, £50 for the gas soldering iron and about £150 for a low voltage temperature controlled iron.

Soldering

There are many ways of making suitable electrical connections and in electrical installation work a screwed terminal is the most common method. In electronics, the most common method of making

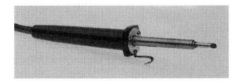

Figure 3.3 Electronic soldering iron and stand.

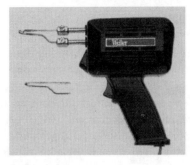

Figure 3.4 Instant soldering gun.

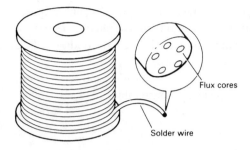

Figure 3.5 Construction of flux cored solder wire.

permanent connections is by soldering the components into the circuit. Good soldering can only be achieved by effort and practice and you should, therefore, take the opportunity to practise the technique before committing your skills to the 'real' situation.

Soldering is an alloying process, whereby a small amount of soft metal (the solder) is made to run between the two metals to be joined, therefore, mixing or alloying them. Solder can be used to join practically any metals or alloys except those containing large amounts of chromium or aluminium, which must be welded or hard soldered.

Soft solders

Soft solders are so called because they are made up of the rather soft metals tin and lead in the proportion 40 to 60. Solders containing tin will adhere very firmly to most metals, providing that the surface of the metals to be joined is clean. Solder will not adhere to a tarnished or oxidised metal surface. This is because solder adheres by forming an alloy with the metal of the connection and this alloy cannot form if there is a film of oxide in the way.

Fluxes

Fluxes are slightly acid materials which dissolve an oxide film, leaving a perfectly clean surface to which the solder can firmly adhere. Rosin fluxes are the most suitable for electrical and electronic work. In electronics it is not convenient to apply the flux and solder separately, so they are combined as flux cored solder wire. This is solder wire with a number of cores of flux running the whole length of the wire. The multicore construction shown in Fig. 3.5 ensures that there is always the correct proportions of flux for each soldered joint.

SOLDERING TECHNIQUES

As already mentioned, when soldering with an iron it is important to choose an iron with a suitable bit size. A 1.5 mm or 2.0 mm bit is suitable for most electronic connections but a 1.0 mm bit is better when soldering dual-in-line (DIL) packages. The bit should be clean and freshly 'tinned'. The materials to be soldered must be free from grease and preferably pre-tinned. Electronic components should not need more cleaning than a wipe to remove dust or grease. The purpose of the soldering iron is to apply heat to the joint. If solder is first melted on to the bit, which is then used to transfer the solder to the joint, the active components of the flux will evaporate before the solder reaches the joint, and an imperfect or 'dry' joint will result. Also applying the iron directly to the joint oxidises the component surfaces, making them more difficult to solder effectively. The best method of making a 'good' soldered joint is to apply the cored solder to the joint and then melt the solder with the iron. This is the most efficient way of heating the termination, letting the solder and the flux carry the heat from the soldering bit on to the termination, as shown in Fig. 3.6.

While the termination is heating up, the solder will appear dull, and then quite suddenly the solder will become bright and fluid, flowing around and 'wetting' the termination. Apply enough solder to cover the termination before removing the solder and then the iron. The joint should be soldered quickly. If attempts are made to improve the joint merely by continued heating and applying more flux and solder, the component or the cable insulation will become damaged by the heat and the connection will have excessive solder on it. The joint must not be moved or blown upon until the solder has solidified. A good soldered joint will appear smooth and bright, a bad

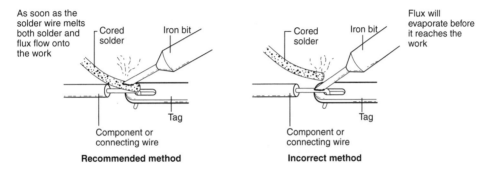

Figure 3.6 Soldering technique with multicore solders.

connection or *dry joint* will appear dull and the solder may be in a 'blob' or appear spiky.

Dry joints

Dry joints may occur because the components or termination are dirty or oxidised, or because the soldering temperature was too low, or too little flux was used. Dry joints do not always make an electrical connection or the connection has a high resistance which deteriorates with time and may cause trouble days or weeks later. A suspected dry joint can be tested as shown in Fig. 3.7. If the joint is 'dry' the voltmeter will read 12 V at position B, just to the right and 0 V at position A to the left of the joint. If the joint is found to be dry. the connection must be remade.

Component assembly and soldering

Soft solder is not as strong as other metals and, therefore, the electronic components must be shaped at the connection site to give extra strength. This can be done by bending the connecting wires so that they hook together or by making the joint area large. Special *lead forming* or *wire shaping* tools are available which both cut and shape the components connecting wires ready for soldering. Figure 3.8 shows a suitable tag terminal connection, Fig. 3.9 a suitable pin terminal connection and Fig. 3.10 a suitable stripboard connection.

All wires must be cut to length before assembly because it is often difficult to trim them after soldering. Also the strain of cutting after soldering may weaken the joint and encourage dry joints. If the wires must be cut after soldering, cutters with a shearing action should be used as shown in Fig. 3.11 Side cutters have a pinching action and the shock of the final pinch-through, identified by a sharp click, may fracture the soldered joint or damage the component.

Most electronic components are very sensitive and are easily damaged by excess heat. Soldered joints must not, therefore, be made close to the body of the component or the heat transferred from the joint may cause some damage. When components are being soldered into a circuit the heat from the soldering iron at the connection must be diverted or 'shunted' away from the body of the component. This can be done by placing a pair of long-nosed pliers or a crocodile clip between the soldered joint and the body of the component as shown in Fig. 3.12.

Components such as resistors, capacitors and transistors are usually cylindrical, rectangular or disc shaped with round wire terminations. They should be shaped, mounted and soldered into the circuit as previously described and shown in Fig. 3.13. A small clearance should be left between the body of the component and the circuit board, to allow convection

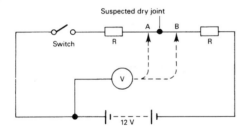

Figure 3.7 Testing a suspected dry joint.

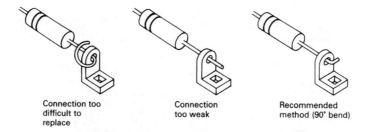

Figure 3.8 Shaping conductors to give strength to electrical connections to tag terminals.

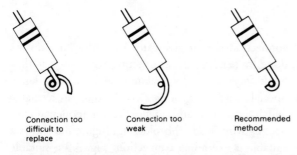

Figure 3.9 Shaping conductors to give strength to electrical connections to pin terminals (plan view).

currents to circulate, which encourages cooling. The vertical mounting method permits many more components to be mounted on the circuit board but the horizontal method gives better mechanical support to the component.

Desoldering

If it is necessary to replace an electronic component, the old, faulty component must first be removed from the circuit board. To do this the solder of the old joint is first liquefied by applying a hot iron to the joint.

The molten solder is then removed from the joint with a desoldering tool. The desoldering tool works like a bicycle pump in reverse and is shown in Fig. 3.14. The tool is made ready by compressing the piston down on to a latch position which holds it closed. The nozzle is then placed into the pool of molten solder and the latch release button pressed. This releases the plunger which shoots out, sucking the molten solder away from the joint and into the body of the desoldering tool.

Removing faulty transistors

First identify the base, collector and emitter connections so that the new component can be correctly connected into the circuit. Remove the solder from each leg with the soldering iron and desoldering tool before removing the faulty transistor. Then, with the aid of a pair of long-nosed pliers, pull the legs of the transistor out of the circuit board. An alternative method is to cut the three legs with a pair of side cutters before desoldering and then remove the individual legs with a pair of long-nosed pliers.

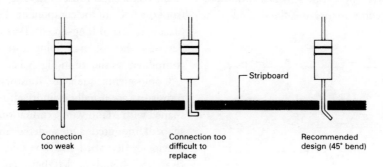

Figure 3.10 Shaping conductors to give strength to electrical connections to stripboard.

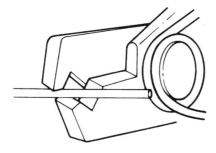

(a) Shearing action

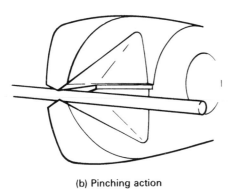

(b) Pinching action

Figure 3.11 Wire cutting.

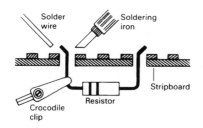

Figure 3.12 Using a crocodile clip as a heat shunt.

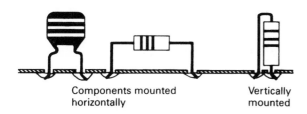

Figure 3.13 Vertical and horizontal mounting of components.

Figure 3.14 A desoldering tool.

Removing faulty integrated circuits (chips)

Remove the solder from each leg of the IC with the soldering iron and desoldering tool and then pull the IC clear of the circuit board. If it has been firmly established that the IC is faulty, it may be removed from the circuit board by cutting the body from the connecting pins before desoldering and removing the individual pins with a pair of long-nosed pliers.

CIRCUIT BOARDS

Permanent circuits require that various discrete components be soldered together on some type of insulated board. Three types of board can be used, matrix, strip and printed circuit board, the base material being plastic synthetic resin bonded paper (SRBP).

Matrix boards

Matrix boards have a matrix of holes on 0.1 inch centres as shown in Fig. 3.15. Boards are available in various sizes; the 149 × 114 mm board is pierced with 58 × 42 holes and the 104 × 65 mm board has 39 × 25 holes. Matrix pins press into any of the holes in the board and provide a terminal post to which components and connecting wires can be soldered. Single sided or double sided matrix pins are available. Double sided pins have the advantage that connections can be made on the underside of the board as well as on the top. The hole spacing of 0.1 inch makes the board compatible with many electronic components. Plug-in relays, DIL integrated circuits and many sockets and connectors all use 0.1 inch spacing at their connections.

Matrix board is probably the easiest and cheapest way to build simple electronic circuits. It is recommended that inexperienced circuit builders construct

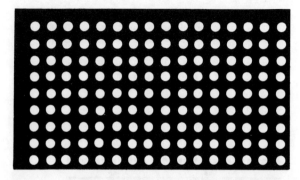
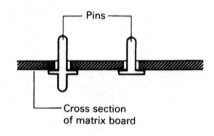

Figure 3.15 Matrix board and double sided and single sided pin inserts.

the circuit on the matrix board using a layout which is very similar to the circuit diagram to reduce the possibility of mistakes.

Suppose, for example, that we intend to build the very simple circuit shown in Fig. 3.16. First we would insert four pins into the matrix board as shown. The diode would then be connected between pins A and B, taking care that the anode was connected to pin A. The resistor would be connected between pins B and C and a wire linked between pins C and D. The a.c. supply from a signal generator would be connected to pins A and D by 'flying' leads and the oscilloscope leads to pins B and C. This circuit would show half wave rectification. When planning the conversion of circuit diagrams into a matrix board layout it helps to have a positional reference system so that we know where to push the pins in the matrix board.

The positional reference system

The positional reference system used with matrix boards uses a simple grid reference system to identify holes on the board. This is achieved by counting along the columns at the top of the board, starting from the left and then counting down the rows. For example, the position reference point 4:3 would be 4 holes from the left and 3 holes down. The board should be prepared as follows:

- turn the matrix board so that a manufactured straight edge is to the top and left-hand side;
- use a felt tip pen to mark the holes in groups of five along the top edge and down the left-hand edge as shown in Fig. 3.17:

The pins can then be inserted as required. Figure 3.17 shows a number of pin reference points. Counting from the left-hand side of the board these are 3:3, 3:16, 10:11, 18:3, 18:11, 25:3 and 25:16.

Stripboard or Veroboard

Stripboard or Veroboard is a matrix board with continuous copper strip attached to one side by adhesive. The copper strips link together rows of holes

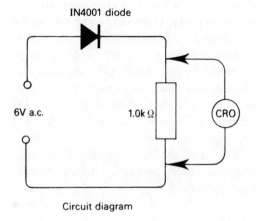

Circuit diagram

Figure 3.16 Circuit diagram converted to a component layout.

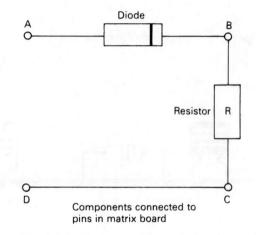

Components connected to pins in matrix board

ELECTRONIC CIRCUIT ASSEMBLY

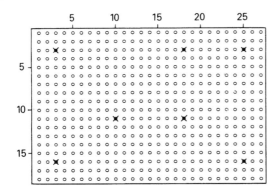

Figure 3.17 Matrix board pin reference system.

so that connections can be made between components inserted into holes on a particular row as shown in Fig. 3.18. The components are assembled on the plain board side with the component leads inserted through the holes and soldered in place to the copper strip.

The copper strips are continuous but they can be broken using a strip cutter or small drill. The drill or cutter is placed on the hole where the break is to be made and then rotated a few turns between the fingers until the very thin copper strip is removed leaving a circle as shown in Fig. 3.18.

Stripboard is very useful because the copper strips take the place of the wire links required with plain matrix boards. Components can easily be mounted vertically on stripboard which leads to high-density small area circuits being assembled. It is, however, more expensive than matrix board because of the additional cost of the copper, most of which is not used in the circuit. Also some translation of the circuit diagram is required before it can be assembled on the stripboard. Excessive heat from the soldering iron can melt the adhesive and cause the copper to peel from the insulating board. Heat should only be applied long enough to melt the solder and secure the component.

Printed circuit boards (PCBs)

Printed circuit boards (PCBs) are produced by chemically etching a copper-clad epoxy glass board so that a copper pattern is engraved on one side of the board. The pattern provides the copper conducting paths making connections to the various components of the circuit. After etching, small holes are drilled for the components which are inserted from the plain side of the board and soldered on to the copper conductor. The copper pattern replaces the lengths of wire used to connect components on to the matrix board.

The copper foil is very thin and is attached to the board with an adhesive. Excessive heat from the soldering iron can melt the adhesive and cause the track to peel from the insulating board. The board should not be flexed, otherwise hairline cracks may appear but go unnoticed until intermittent faults occur in the circuit later.

The process of designing and making a PCB is quite simple but does require specialised equipment and for that reason we will not consider it further here.

Wire wrapping

As the electronic industry increasingly uses advanced technology, the demand for a faster, more reliable and inexpensive method of making electrical connections

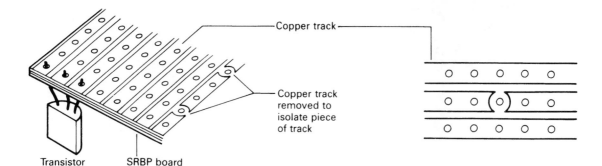

Figure 3.18 Stripboard or Veroboard.

has increased. Electronic equipment today has lots of components with many terminals in a very small space. To make electrical connections to these high-density electronic circuits, the electronic industry has developed a new solderless wire wrapping technique.

A wire wrapped connection is made by winding a special insulated wire of 30 AWG (0.25 mm) around the sharp corners of a square pin inserted into the circuit board. The winding tension causes the corners of the pin to cut into the wire, producing a good electrical connection which will not unwind and is as good as or better than a soldered joint. This method of connection was developed by the Bell Telephone Laboratories of the Western Electric Company.

Pneumatic or electric tools are preferred for production work but battery- or hand-operated tools as shown in Fig. 3.19 are used for service and repair work.

To make the connection:

- the end of the wire is inserted into the bit of the tool and then bent back at 90° to the tool;
- the bit is placed over the terminal pin;
- the tool is rotated clockwise a few turns without undue pressure;
- the bit is removed from the terminal pin;
- the connection is made.

One advantage of wire wrapping is the ease with which a wire can be removed from a terminal because of an error or wiring modification. An unwrap tool is placed over the terminal pin, rotated anti-clockwise, and the connection is removed in seconds without damage to the terminal pin.

Wire wrapping is a precision technique and the bit size and wire diameter must be compatible with the terminal pin size if a good electrical connection is to be made.

Wire wrapping as a method of making electronic connections has not had the universal appeal it was once thought it might.

Breadboards or prototype

Breadboarding is the name given to solderless temporary circuit building by pressing wires and component leads into holes in the prototype board. This method is used for building temporary circuits for testing or investigation.

The S-DeC

The S-DeC prototype board is designed for interconnecting discrete components. The hole spacing and hole connections do not suit DIL IC packages. Each board has 70 phosphor-bronze contact points arranged in two sections, each of which has seven parallel rows of five connected contact points. The case is formed in high impact polystyrene and the individual boards may be interlocked to create a larger working area. The S-DeC can be supplied with a vertical bracket for mounting switches or variable resistors as shown in Fig. 3.20

The professional

The professional prototype board is designed for the interconnection of many different types of component. The hole spacing of 0.1 inch allows DIL IC packages to be plugged directly into the board. Each board has 47 rows of five interconnected nickel-silver

battery operated

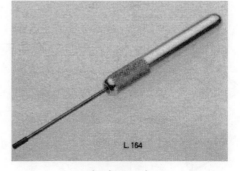

hand operated

Figure 3.19 Wire wrapping tools.

Figure 3.20 S-DeC prototype board used for temporary circuit building.

contacts each side of a central channel and a continuous row at the top and bottom which may be used as power supply rails. The case is formed in high impact thermoplastic and the individual boards may be interlocked to create a larger working area. A vertical side bracket is also supplied for mounting switches or variable resistors as shown in Fig. 3.21.

Interconnection methods

A plug and socket provide an ideal method of connecting or isolating components and equipment which cannot be permanently connected. In electrical installation work we usually need to make plug and socket connections between three conductors on single phase circuits and five conductors on three phase systems. In electronics we often need to make multiple connections between circuit boards or equipment. However, the same principles apply, that is the plug and socket must be capable of separation, but while connected they must make a good electrical contact. Also the plug and socket must incorporate some method of preventing reverse connection.

PCB EDGE CONNECTORS

A range of connectors are available which make direct contact to printed circuit boards as shown in Fig. 3.22. Multiple connectors are available with a contact pitch of 0.1 inch so that they can be soldered into circuit boards. The plug and socket can then be used as edge connectors to make board-to-board and cable-to-board connections.

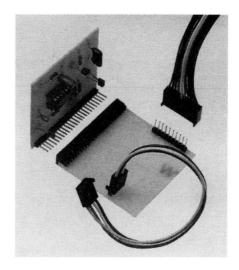

Figure 3.22 PCB edge connectors.

RIBBON CABLE CONNECTORS

A ribbon cable is a multicore cable laid out as flat strip or ribbon strip. A range of connectors are made to connect ribbon cable, as shown in Fig. 3.23, which is used for making board-to-board interconnections and to connect computer peripherals such as VDUs and printers.

DIN CONNECTORS

DIN-style audio connectors are available for making up to eight connections as shown in Fig. 3.24 and used when frequent connection and disconnection are required between a small number of contacts.

JACK CONNECTORS

Figure 3.21 Professional prototype board used for temporary circuit building.

Jack connectors are used when frequent connections are to be made between two or three poles on, for

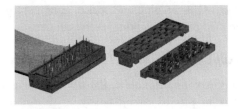

Figure 3.23 Ribbon cable connector.

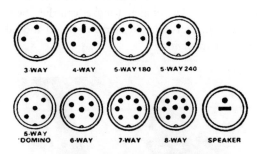

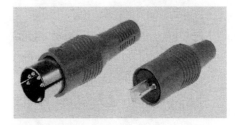

Figure 3.24 'DIN'-style audio connectors.

example, headphones or microphones. They are available in three sizes, sub-miniature (2.5 mm), miniature (3.5 mm) and commercial (0.25 inch) as shown in Fig. 3.25.

All the above connectors may be terminated onto the cable end by a soldered, crimped or cable displacement method. When a soldered connection is to be made the cable end must be stripped of its insulation, tinned and then terminated. A crimped connection also requires that the insulation be removed and the prepared cable end inserted into a lug, which is then crimped using an appropriate tool. The insulation displacement method of connection is much quicker to make because the cable ends do not require stripping or preparing. The connection is made by pressing insulation piercing tines or prongs into the cable which displaces the insulation to make an electrical connection with the conductor. This method is used extensively when terminating ribbon cable and for making rapid connections to the existing wiring system of motor vehicles.

Figure 3.25 Jack connectors.

Fault finding

The best way to avoid problems in electronic circuit assembly is to be always alert while working, to think about what you are doing and always try to be neat. If, despite your best efforts, the circuit does not work as it should when tested, then follow a logical test procedure which will usually find the fault in the shortest possible time. First carry out a series of visual tests.

1. Is the battery or supply correctly connected?
2. Is the battery flat or the supply switched on?
3. Is the circuit constructed *exactly* as it should be according to the circuit diagram?
4. Are all the components in place?
5. Check the values of all the components.
6. Are all the components such as diodes, capacitors, transistors and ICs connected the correct way round?
7. Have all connections and links been made?
8. Have all the necessary breaks been made in the stripboard, e.g. between the IC?
9. Are all the soldered joints good?
10. Are any of the components hot or burnt?

If the fault has not been identified by the first ten tests, ask someone else to carry them out. You may have missed something which will be obvious to someone else. If the visual tests have failed to identify the fault, then further meter tests are called for.

11. Check the input voltage and the output voltage. Check the mid-point voltage between components which are connected in series with the supply.
12. Variable resistors may suffer from mechanical wear. Check the voltage at the wiper as well as across the potentiometer.
13. Check the coil voltage on relays; if this is low, the coil contacts may not be making.
14. Is the diode connected correctly? Short circuit the

diode momentarily with a wire link to see if the circuit works. If it does the diode is open circuit.
15 Check resistor capacitor circuits by momentarily shorting out the capacitor and then observing the charging voltage. If it does not charge, the resistor may be open circuit. If it charges instantly, the resistor may be short circuit. Check the polarity of electrolytic capacitors. Check the capacitor leads for breaks where the lead enters the capacitor body.
16 Check the base-emitter voltage of the transistor. A satisfactory reading would be between 0.6 V and 1.0 V. Temporarily connect a 1 kΩ resistor between the positive supply and the base connection. If the transistor works, the base feed is faulty. If it does not work, the transistor is faulty.
17 Short out the anode and cathode of the thyristor. If the load operates, the thyristor or the gate pulse is faulty. If the load does not operate, the load is faulty.

The testing of capacitors and resistors is further discussed in Chapter 2, while the testing of discrete semiconductor components is dealt with in Chapter 4.

4

SEMICONDUCTOR DEVICES

Semiconductor materials

Modern electronic devices use the semiconductor properties of materials such as silicon or germanium. The atoms of pure silicon or germanium are arranged in a lattice structure as shown in Fig. 4.1. The outer electron orbits contain four electrons known as valence electrons. These electrons are all linked to other valence electrons from adjacent atoms forming a co-valent bond. There are no free electrons in pure silicon or germanium and, therefore, no conduction can take place unless the bonds are broken and the lattice framework is destroyed.

To make conduction possible without destroying the crystal it is necessary to replace a four-valent atom with a three- or five-valent atom. This process is known as *doping*.

If a three-valent atom is added to silicon or germanium a hole is left in the lattice framework. Since the material has lost a negative charge, the material becomes positive and is known as a p-type material, p for positive.

If a five-valent atom is added to silicon or germanium, only four of the valence electrons can form a bond and one electron becomes mobile or free to carry charge. Since the material has gained a negative charge it is known as an n-type material, n for negative.

Bringing together a p-type and n-type material allows current to flow in one direction only through the p-n junction. Such a junction is called a diode since it is the semiconductor equivalent of the vacuum diode valve used by Fleming to rectify radio signals in 1904.

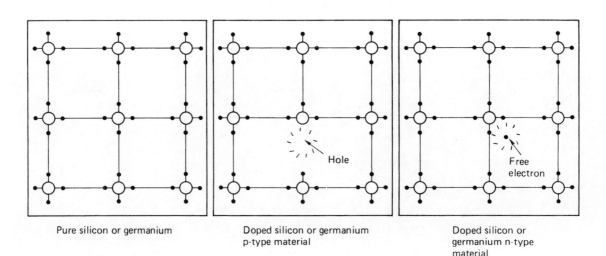

Figure 4.1 Semiconductor material.

Semiconductor diode

A semiconductor or junction diode consists of a p-type and n-type material formed in the same piece of silicon or germanium. The p-type material forms the anode and the n-type the cathode as shown in Fig. 4.2. If the anode is made positive with respect to the cathode, the junction will have very little resistance and current will flow. This is referred to as forward bias. However, if reverse bias is applied, that is, the anode is made negative with respect to the cathode, the junction resistance is high and no current can flow as shown in Fig. 4.3. The characteristics for a forward and reverse bias p-n junction are given in Fig. 4.4.

It can be seen that a small voltage is required to forward bias the junction before a current can flow. This is approximately 0.6 V for silicon and 0.2 V for germanium. The reverse bias potential of silicon is about 1200 V and for germanium about 300 V. If the reverse bias voltage is exceeded the diode will break down and current will flow in both directions. Similarly, the diode will break down if the current rating is exceeded, because excessive heat will be generated. Manufacturers' information therefore gives maximum voltage and current ratings for individual diodes which must not be exceeded (see Appendix F). However, it is possible to connect a number of standard diodes in series or parallel, thereby sharing current or voltage as shown in Fig. 4.5 so that the manufacturers' maximum values are not exceeded by the circuit.

DIODE TESTING

The p-n junction of the diode has a low resistance in one direction and a very high resistance in the reverse direction.

Connecting an ohm meter, as described in Chapter 6, with the red positive lead to the anode of the junction diode and the black negative lead to the cathode, would give a very low reading. Reversing the lead connections would give a high resistance reading in a 'good' component.

ZENER DIODE

A zener diode is a silicon junction diode but with a different characteristic to the semiconductor diode considered previously. It is a special diode with a pre-determined reverse breakdown voltage, the

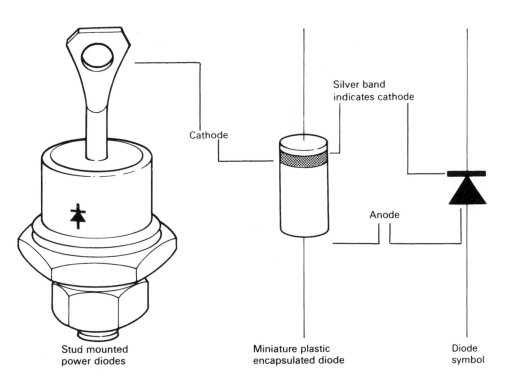

Figure 4.2 Symbol and appearance of semiconductor diodes.

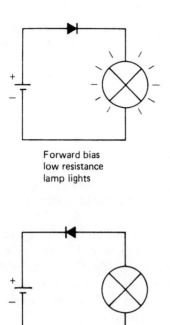

Figure 4.3 Forward and reverse bias of a diode.

mechanism for which was discovered by Carl Zener in 1934. The symbol and general appearance are shown in Fig. 4.6. In its forward bias mode, that is when the anode is positive and the cathode negative, the zener will conduct at about 0.6 V, just like an ordinary diode, but it is in the reverse mode that the zener diode is normally used. When connected with the anode made negative and the cathode positive, the reverse current is zero until the reverse voltage reaches a pre-determined value when the diode switches on as shown by the characteristics given in Fig. 4.7. This is called the zener voltage or reference voltage. Zener diodes are manufactured in a range of preferred values, for example, 2.7 V, 4.7 V, 5.1 V, 6.2 V, 6.8 V, 9.1 V, 10 V, 11 V, 12 V etc., up to 200 V at various ratings, as shown by the tables in appendix F. The diode may be damaged by overheating if the current is not limited by a series resistor, but when this is connected, the voltage across the diode remains constant. It is this property of the zener diode which makes it useful for stabilising power supplies and these circuits are considered in Chapter 5.

If a test circuit is constructed as shown in Fig. 4.8

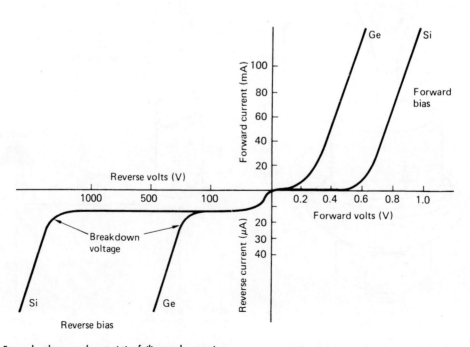

Figure 4.4 Forward and reverse characteristic of silicon and germanium.

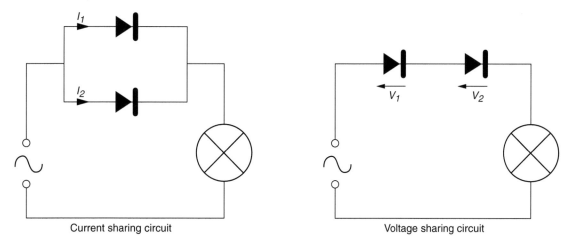

Figure 4.5 Using two diodes to reduce the current or voltage applied to a diode.

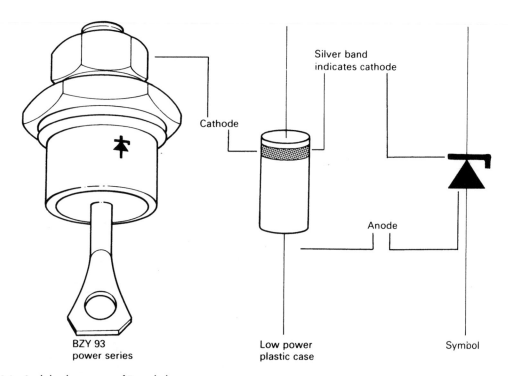

Figure 4.6 Symbol and appearance of Zener diodes.

the zener action can be observed. When the supply is less than the zener voltage (5.1 V in this case) no current will flow and the output voltage will be equal to the input voltage. When the supply is equal to or greater than the zener voltage, the diode will conduct and any excess voltage will appear across the 680 Ω resistor resulting in a very stable voltage at the output. When connecting this and other electronic circuits you must take care to connect the polarity of the zener diode as shown in the diagram. Note that current must flow through the diode to enable it to stabilise.

LIGHT-EMITTING DIODE (LED)

The light-emitting diode is a p-n junction especially manufactured from a semiconducting material which

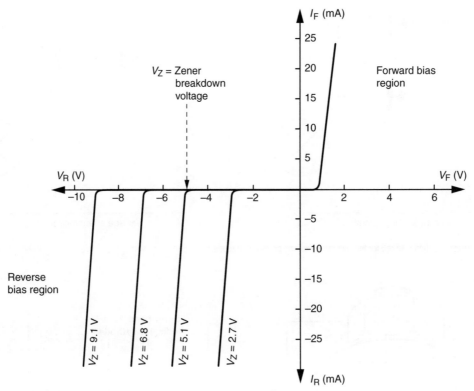

Figure 4.7 Zener diode characteristics.

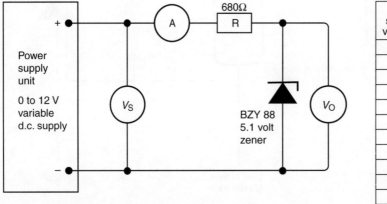

PSU supply volts V_S	Current A	Output volts V_O
1		
2		
3		
4		
5		
6		
7		
8		
9		
10		
11		
12		

Figure 4.8 Experiment to demonstrate the operation of a Zener diode.

emits light when a current of about 10 mA flows through the junction.

No light is emitted when the junction is reverse biased and if this exceeds about 5 V the LED may be damaged.

The general appearance and circuit symbol are shown in Fig. 4.9.

The LED will emit light if the voltage across it is about 2 V. If a voltage greater than 2 V is to be used then a resistor must be connected in series with the LED.

To calculate the value of the series resistor we must ask ourselves what we know about LEDs. We know that the diode requires a forward voltage of

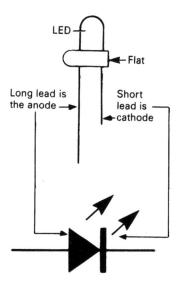

Figure 4.9 Symbol and general appearance of an LED.

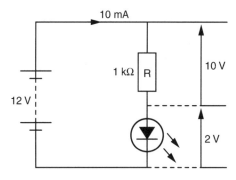

Figure 4.10 Circuit diagram for LED example.

about 2 V and a current of about 10 mA must flow through the junction to give sufficient light. The value of the series resistor R will, therefore, be given by

$$R = \frac{\text{supply voltage} - 2\text{ V}}{10\text{ mA}}\ \Omega$$

EXAMPLE 1

Calculate the value of the series resistor required when an LED is to be used to show the presence of a 12 V supply.

$$R = \frac{12\text{ V} - 2\text{ V}}{10\text{ mA}}\ \Omega$$

$$= \frac{10\text{ V}}{10\text{ mA}} = 1\text{ k}\Omega$$

The circuit is, therefore, as shown in Fig. 4.10.

LEDs are available in red, yellow and green and when used with a series resistor may replace a filament lamp. They use less current than a filament lamp, are smaller, do not become hot and last indefinitely. A filament lamp, however, is brighter and emits white light. LEDs are often used as indicator lamps, to indicate the presence of a voltage. They do not, however, indicate the *precise* amount of voltage present at that point.

Another application of the LED is the seven-segment display used as a numerical indicator in calculators, digital watches and measuring instruments. Seven LEDs are arranged as a figure eight so that when various segments are illuminated, the numbers 0 to 9 are displayed as shown in Fig. 4.11.

Light-dependent resistor (LDR)

Almost all materials change their resistance with a change in temperature. Light energy falling on a suitable semiconductor material also causes a change in resistance. The semiconductor material of an LDR is encapsulated as shown in Fig. 4.12 together with the circuit symbol. The resistance of an LDR in total darkness is about 10 MΩ, in normal room lighting about 5 kΩ and in bright sunlight about 100 Ω. They can carry tens of milli-amperes, an amount which is sufficient to operate a relay. The LDR uses this characteristic to switch on automatically street lighting and security alarms.

PHOTODIODE

The photodiode is a normal junction diode with a transparent window through which light can enter. The circuit symbol and general appearance are shown in Fig. 4.13. It is operated in reverse bias mode and the leakage current increases in proportion to the amount of light falling on the junction. This is due to the light energy breaking bonds in the crystal lattice of the semiconductor material to produce holes and electrons.

Photodiodes will only carry micro-amperes of current but can operate much more quickly than LDRs and are used as 'fast' counters when the light intensity is changing rapidly.

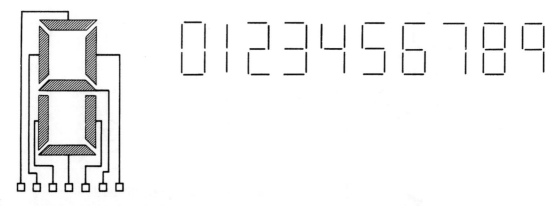

Figure 4.11 LED used in seven-segment display.

Figure 4.12 Symbol and appearance of a light-dependent resistor.

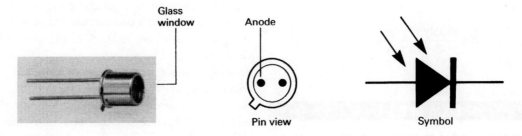

Figure 4.13 Symbol, pin connections and appearance of a photodiode.

Thermistor

The thermistor is a thermal resistor, a semiconductor device whose resistance varies with temperature. The circuit symbol and general appearance are shown in Fig. 4.14. They can be supplied in many shapes and are used for the measurement and control of temperature up to their maximum useful temperature limit of about 300°C. They are very sensitive and because the bead of semiconductor material can be made very small, they can measure temperature in the most inaccessible places with very fast response times. Thermistors are embedded in high-voltage underground transmission cables in order to monitor the temperature of the cable. Information about the temperature of a cable allows engineers to load the cables more efficiently. A particular cable can carry a larger load in winter for example, when heat from the cable is being dissipated more efficiently. A thermistor is also used to monitor the water temperature of a motor car.

Transistors

The transistor has become the most important single building block in electronics. It is the modern,

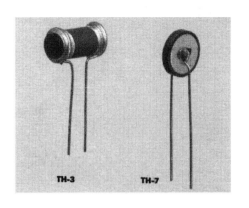 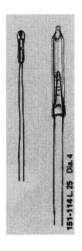 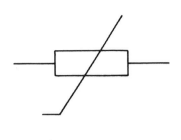

Figure 4.14 Symbol and appearance of a thermistor.

miniature, semiconductor equivalent of the thermionic valve and was invented in 1947 by Bardeen, Shockley and Brattain at the Bell Telephone Laboratories in the USA. Transistors are packaged as separate or *discrete* components as shown in Fig. 4.15.

There are two basic types of transistor, the *bipolar* or junction transistor and the *field effect transistor* (FET).

The FET has some characteristics which make it a better choice in electronic switches and amplifiers. It uses less power and has a higher resistance and frequency response. It takes up less space than a bipolar transistor and, therefore, more of them can be packed together on a given area of silicon chip. It is, therefore, the FET which is used when many transistors are integrated on to a small area of silicon chip as in the *integrated circuit* (IC) discussed later.

When packaged as a discrete component the FET looks much the same as the bipolar transistor. The circuit symbol and connections are given in Appendix H. However, it is the bipolar transistor which is much more widely used in electronic circuits as a discrete component.

THE BIPOLAR TRANSISTOR

The bipolar transistor consists of three pieces of semiconductor material sandwiched together as shown in Fig. 4.16. The structure of this transistor makes it a three-terminal device having a base, collector and emitter terminal. By varying the current flowing into the base connection a much larger current flowing between collector and emitter can be controlled. Apart from the supply connections, both n-p-n and

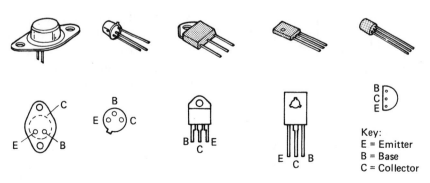

Transistor families

Figure 4.15 The appearance and pin connections of the transistor family.

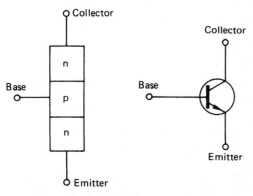 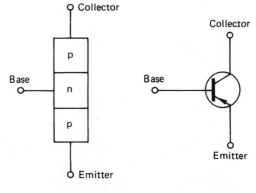

(a) Structure and symbol of n-p-n transistor (b) Structure and symbol of p-n-p transistor

Figure 4.16 Structure and symbol of n-p-n and p-n-p transistors.

p-n-p types are essentially the same but the n-p-n type is more common.

A transistor is generally considered a current-operated device. There are two possible current paths through the transistor circuit, shown in Fig. 4.17, the base-emitter path when the switch is closed and the collector-emitter path. Initially, the positive battery supply is connected to the n-type material of the collector, the junction is reverse biased and, therefore, no current will flow. Closing the switch will forward bias the base-emitter junction and current flowing through this junction causes current to flow across the collector-emitter junction and the signal lamp will light.

A small base current can cause a much larger collector current to flow. This is called the *current gain* of the transistor and is typically about 100. When I say a much larger collector current, I mean a large current in electronic terms, up to about half an ampere.

We can, therefore, regard the transistor as operating in two ways, as a switch because the base current turns on and controls the collector current, and secondly as a current amplifier because the collector current is greater than the base current.

We could also consider the transistor to be operating in a similar way to a relay. However, the transistor has many advantages over electrically operated switches such as relays. They are very small, reliable, have no moving parts and, in particular, they can switch millions of times a second without arcing occurring at the contacts.

TRANSISTOR TESTING

A transistor can be thought of as two diodes connected together and, therefore, a transistor can be tested using an ohm meter in the same way as was described for the diode.

Assuming that the red lead of the ohm meter is positive, as described in Chapter 6, the transistor can be tested in accordance with Table 4.1.

When many transistors are to be tested, a simple test circuit can be assembled as shown in Fig. 4.18.

With the circuit connected, as shown in Fig. 4.18, a 'good' transistor will give readings on the voltmeter of 6 V with the switch open and about 0.5 V when the switch is made. The voltmeter used for the test should have a high internal resistance, about ten times greater than 4.7 kΩ. This is usually indicated on the back of a multi-range meter or in the manufacturers' information supplied with a new meter.

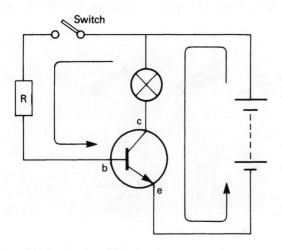

Figure 4.17 Operation of the transistor.

Table 4.1 Transistor testing using an ohm meter

A 'good' n-p-n transistor will give the following readings

Red to base and black to collector = low resistance
Red to base and black to emitter = low resistance

Reversed connections on the above terminals will result in a high resistance reading as will connections of either polarity between the collector and emitter terminals.

A 'good' p-n-p transistor will give the following readings

Black to base and red to collector = low resistance
Black to base and red to emitter = low resistance

Reversed connections on the above terminals will result in a high resistance reading as will connections of either polarity between the collector and emitter terminals.

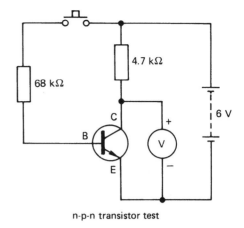

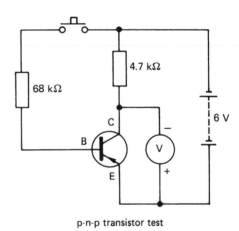

Figure 4.18 Transistor test circuits.

Integrated circuits

Integrated circuits were first developed in the 1960s. They are densely populated miniature electronic circuits made up of hundreds and sometimes thousands of microscopically small transistors, resistors, diodes and capacitors, all connected together on a single chip of silicon no bigger than a baby's little finger nail. When assembled in a single package, as shown in Fig. 4.19, we call the device an integrated circuit (IC).

There are two broad groups of integrated circuit, digital ICs and linear ICs. Digital ICs contain simple switching-type circuits used for logic control and calculators, discussed in Chapter 7. Linear ICs incorporate amplifier-type circuits which can respond to audio and radio frequency signals. The most versatile linear IC is the operational amplifier which has applications in electronics, instrumentation and control, and its application to strain gauges is discussed in Chapter 12.

The integrated circuit is the electronic revolution. ICs are more reliable, cheaper, smaller and electronically superior to the same circuit made from discrete or separate transistors. One IC behaves differently to another because of the arrangement of the transistors within the IC.

Manufacturers' data sheets describe the characteristics of the different ICs which have a reference number stamped on the top. Appendix L gives the characteristics of some of the more common ICs.

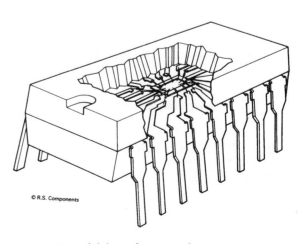

Figure 4.19 Exploded view of an integrated circuit.

When building circuits, it is necessary to be able to identify the IC pin connection by number. The number one pin of any IC is indicated by a dot pressed into the encapsulation or, the number one pin is the pin to the left of the cutout as shown in Fig. 4.20. Since the packaging of ICs has two rows of pins they are called DIL (dual in line) packaged integrated circuits and their appearance is shown in Fig. 4.21.

Integrated circuits are sometimes connected into DIL sockets and at other times are soldered directly into the circuit. The testing of ICs is beyond the scope of a practising electrician and when they are suspected of being faulty an identical or equivalent replacement should be connected into the circuit, ensuring that it is inserted the correct way round, which is indicated by the position of pin number one as described earlier.

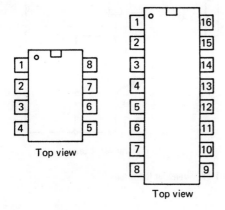

Figure 4.20 IC pin identification.

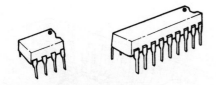

Figure 4.21 DIL packaged integrated circuits.

The thyristor or silicon-controlled rectifier (SCR)

The thyristor was previously known as a *silicon controlled rectifier* since it is a rectifier which controls the power to a load. It consists of four pieces of semiconductor material sandwiched together and connected to three terminals as shown in Fig. 4.22.

The word thyristor is derived from the Greek word *thyra* meaning door, because the thyristor behaves like a door. It can be open or shut, allowing or preventing current flow through the device. The door is opened, or we say the thyristor is triggered, to a conducting state by applying a pulse voltage to the gate connection. Once the thyristor is in the conducting state, the gate loses all control over the device. The only way to bring the thyristor back to a non-conducting state is to reduce the voltage across the anode and cathode to zero or apply reverse voltage across the anode and cathode.

We can understand the operation of a thyristor by considering the circuit shown in Fig. 4.23. This circuit can also be used to test suspected faulty components.

THYRISTOR OPERATION AND TESTING

In Fig. 4.23, when SWB only is closed the lamp will not light, but when SWA is also closed, the lamp lights to full brilliance. The lamp will remain illuminated even when SWA is opened. This shows that the thyristor is operating correctly. Once a voltage has been applied to the gate the thyristor becomes forward conducting, like a diode, and the gate loses control.

THYRISTOR IN PRACTICE

The thyristor has no moving parts and operates without arcing. It can operate at extremely high speeds and the currents used to operate the gate are very small. The most common application for the thyristor is to control the power supply to a load, for example, lighting dimmers and motor speed control. A number of circuits are considered in Chapter 5.

The power available to an a.c. load can be controlled by allowing current to be supplied to the load during only a part of each cycle. This can be achieved by supplying a gate pulse automatically at a chosen point in each cycle as shown by Fig. 4.24. Power is reduced by triggering the gate later in the cycle.

The thyristor is only a half-wave device (like a diode) allowing control of only half the available power in an a.c. circuit. This is very uneconomical and a further development of this device has been the triac, which is considered next.

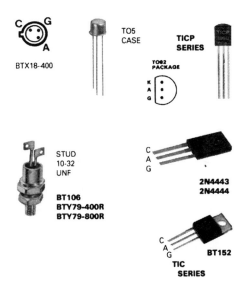
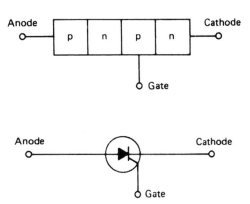

Figure 4.22 Structure, symbol and appearance of a thyristor.

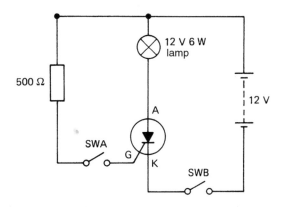

Figure 4.23 Thyristor test circuit.

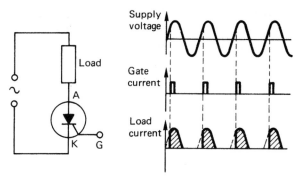

Figure 4.24 Waveforms to show the control effect of a thyristor.

THYRISTOR TESTING USING AN OHM METER

A thyristor may also be tested using an ohm meter as described in Table 4.2 assuming that the red lead of the ohm meter is positive as described in Chapter 6.

The triac

The triac was developed following the practical problems experienced in connecting two thyristors in parallel, to obtain full wave control, and in providing two separate gate pulses to trigger the two devices.

The triac is a single device containing a back-to-back, two-directional thyristor which is triggered on both halves of each cycle of the a.c. supply by the same gate signal. The power available to the load can, therefore, be varied between zero and full load.

The symbol and general appearance are shown in Fig. 4.25. Power to the load is reduced by triggering the gate later in the cycle as shown by the waveforms of Fig. 4.26.

Table 4.2 Thyristor testing using an ohm meter

A 'good' thyristor will give the following readings

Black to cathode and red on gate = low resistance
Red to cathode and black on gate = a higher resistance value

The value of the second reading will depend upon the thyristor and may vary from only slightly greater to very much greater.

From cathode to anode with either polarity connected will result in a very high resistance reading.

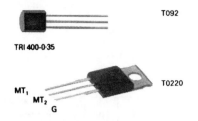

Figure 4.25 Appearance of a triac.

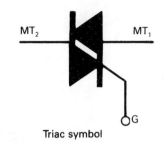

Triac symbol

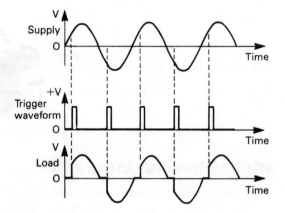

Figure 4.26 Waveforms to show the control effect of a triac.

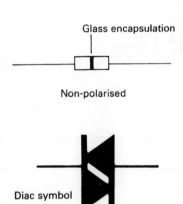

Figure 4.27 The symbol and appearance of a diac used in triac firing circuits.

The triac is a three-terminal device, just like the thyristor, but the terms anode and cathode have no meaning for a triac. Instead, they are called main terminal one (MT1) and main terminal two (MT2). The device is triggered by applying a small pulse to the gate (G). A gate current of 50 mA is sufficient to trigger a triac switching up to 100 A. They are used for many commercial applications where control of a.c. power is required, for example, motor speed control and lamp dimming.

The diac

The diac is a two-terminal device containing a two-directional zener diode. It is used mainly as a trigger device for the thyristor and triac. The symbol is shown in Fig. 4.27.

The device turns on when some predetermined voltage level is reached, say 30 V, and, therefore, it can be used to trigger the gate of a triac or thyristor each time the input waveform reaches this predetermined value. Since the device contains back-to-back zener dodes it triggers on both the positive and negative half cycles.

The applications of these semiconductor devices in various electronic circuits are considered in Chapter 5.

Manufacturers' data sheet information and practical information on these semiconductor devices is given in Appendices E, F, G, H, I and J.

5

ELECTRONIC CIRCUITS IN ACTION

Voltage divider

In Chapter 8 we consider the distribution of voltage across resistors connected in series. We find that the supply voltage is divided between the series resistors in proportion to the size of the resistor. If two identical resistors were connected in series across a 12 V supply as shown in Fig. 5.1(a) both common sense and a simple calculation would confirm that 6 V would be measured across the output. In the circuit shown in Fig. 5.1(b) the 1 kΩ and 2 kΩ resistors divide the input voltage into three equal parts. One part, 4 V, will appear across the 1 kΩ resistor and two parts, 8 V, will appear across the 2 kΩ resistor. In Fig. 5.1(c) the situation is reversed and, therefore, the voltmeter will read 4 V. The division of the voltage is proportional to the ratio of the two resistors and, therefore, we call this simple circuit a voltage divider or potential divider. The values of the resistors R_1 and R_2 determine the output voltage as follows:

$$V_{OUT} = V_{IN} \times \frac{R_2}{R_1 + R_2} \text{ Volts}$$

For the circuit shown in Fig. 5.1(b)

$$V_{OUT} = 12 \text{ V} \times \frac{2 \text{ k}\Omega}{1 \text{ k}\Omega + 2 \text{ k}\Omega} = 8 \text{ V}$$

For the circuit shown in Fig. 5.1(c)

$$V_{OUT} = 12 \text{ V} \times \frac{1 \text{ k}\Omega}{2 \text{ k}\Omega + 1 \text{ k}\Omega} = 4 \text{ V}$$

EXAMPLE 1

For the circuit shown in Fig. 5.2 calculate the output voltage

$$V_{OUT} = 6 \text{ V} \times \frac{2.2 \text{ k}\Omega}{10 \text{ k}\Omega + 2.2 \text{ k}\Omega} = 1.08 \text{ V}$$

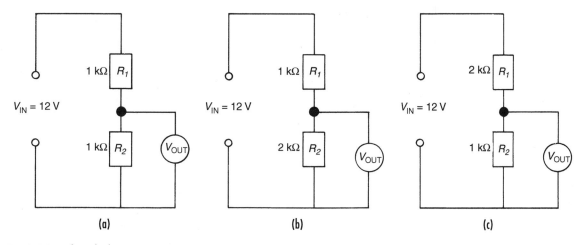

Figure 5.1 Voltage divider circuit.

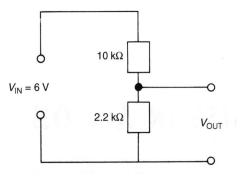

Figure 5.2 Voltage divider circuit for Example 1.

EXAMPLE 2

For the circuit shown in Fig. 5.3(a) calculate the output voltage. We must first calculate the equivalent resistance of the parallel branch

$$\frac{1}{R_T} = \frac{1}{R_1} + \frac{1}{R_2}$$

$$\frac{1}{R_T} = \frac{1}{10\,k\Omega} + \frac{1}{10\,k\Omega} = \frac{1+1}{10\,k\Omega} = \frac{2}{10\,k\Omega}$$

$$R_T = \frac{10\,k\Omega}{2} = 5\,k\Omega$$

The circuit may now be considered as shown in Fig. 5.3(b)

$$V_{OUT} = 6V \times \frac{10\,k\Omega}{5\,k\Omega + 10\,k\Omega} = 4\,V$$

Voltage dividers are used in electronic circuits to produce a reference voltage which is suitable for operating transistors and integrated circuits. The volume control in a radio or the brightness control of a CRO requires a continuously variable voltage divider and this can be achieved by connecting a variable resistor or potentiometer as shown in Fig. 5.4. With the wiper arm making a connection at the bottom of the resistor, the output would be 0 V. When connection is made at the centre, the voltage would be 6 V, and at the top of the resistor the voltage would be 12 V. The voltage is continuously variable between 0 V and 12 V simply by moving the wiper arm of a suitable variable resistor such as those shown in Fig. 2.3.

When a load is connected to a voltage divider it 'loads' the circuit, causing the output voltage to fall below the calculated value. To avoid this, the resistance of the load should be at least ten times greater than the value of the resistor across which it is connected. For example, the load connected across the voltage divider shown in Fig. 5.1(b) must be greater than 20 kΩ and across 5.1(c) greater than 10 kΩ. This problem of loading the circuit also occurs when taking voltage readings as we will discuss in Chapter 6.

Rectification of a.c.

When a d.c. supply is required, batteries or a rectified a.c. supply can be provided. Batteries have the advantage of portability but a battery supply is more expensive than using the a.c. mains supply suitably

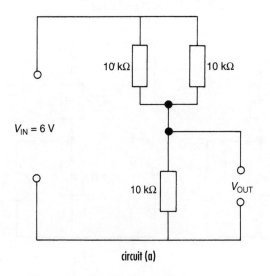

circuit (a)

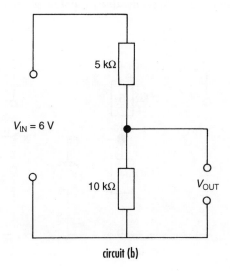

circuit (b)

Figure 5.3 Voltage divider circuit for Example 2.

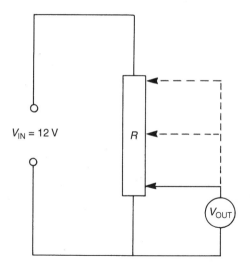

Figure 5.4 Constantly variable voltage divider circuit.

rectified. Rectification is the conversion of an alternating current supply into a uni-directional or direct current supply. This is one of the many applications for a diode which will conduct in one direction only, that is when the anode is positive with respect to the cathode as we discussed in Chapter 4 under Semiconductor diode.

HALF-WAVE RECTIFICATION

The circuit is connected as shown in Fig. 5.5. During the first half cycle the anode is positive with respect to the cathode and, therefore, the diode will conduct. When the supply goes negative during the second half cycle, the anode is negative with respect to the cathode and, therefore, the diode will not allow current to flow. Only the positive half of the waveform will be available at the load and the lamp will light at reduced brightness.

FULL-WAVE RECTIFICATION

Figure 5.6 shows an improved rectifier circuit which makes use of the whole a.c. waveform and is, therefore, known as a full-wave rectifier. When the four diodes are assembled in this diamond-shaped configuration, the circuit is also known as a *bridge rectifier*. During the first half cycle diodes D_1 and D_3 conduct and diodes D_2 and D_4 conduct during the second half cycle. The lamp will light to full brightness.

Full-wave and half-wave rectification can be displayed on the screen of a CRO and will appear as shown in Figs. 5.5 and 5.6.

Smoothing

The circuits of Figs. 5.5 and 5.6 convert an alternating waveform into a waveform which never goes negative, but they cannot be called continuous d.c. because they contain a large alternating component. Such a waveform is too bumpy to be used to supply electronic equipment but may be used for battery charging. To be useful in electronic circuits the output must be smoothed. The simplest way to smooth an output is to connect a large-value capacitor across the output terminals as shown in Fig. 5.7.

When the output from the rectifier is increasing, as shown by the dotted lines of Fig. 5.8, the capacitor

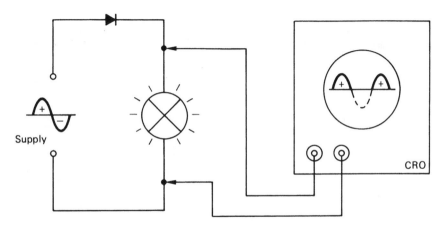

Figure 5.5 Half-wave rectification.

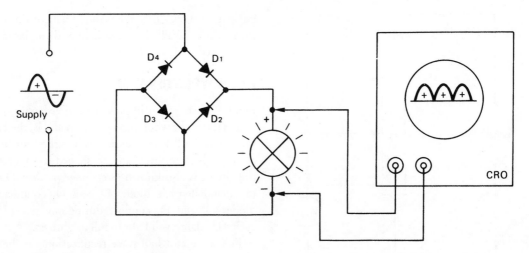

Figure 5.6 Full-wave rectification using a bridge circuit.

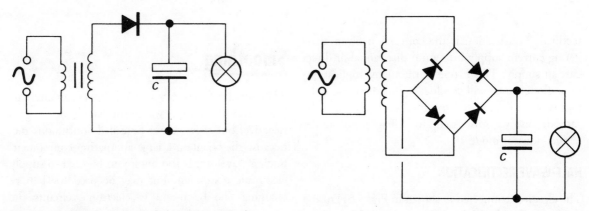

Figure 5.7 Rectified a.c. with smoothing capacitor connected.

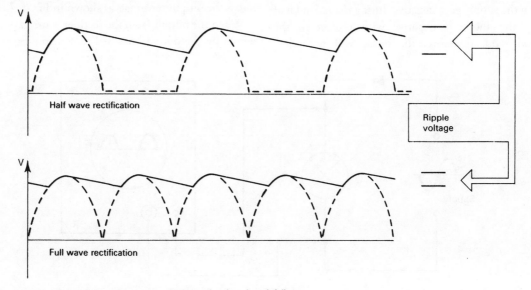

Figure 5.8 Output waveforms with smoothing showing reduced ripple with fullwave.

charges up. During the second quarter of the cycle, when the output from the rectifier is falling to zero, the capacitor discharges into the load. The output voltage falls until the output from the rectifier once again charges the capacitor. The capacitor connected to the full-wave rectifier circuit is charged up twice as often as the capacitor connected to the half-wave circuit and, therefore, the output ripple on the full-wave circuit is smaller, giving better smoothing. Increasing the current drawn from the supply increases the size of the ripple. Increasing the size of the capacitor reduces the amount of ripple.

LOW PASS FILTER

The ripple voltage of the rectified and smoothed circuit shown in Fig. 5.7 can be further reduced by adding a low pass filter, as shown in Fig. 5.9. A low pass filter allows low frequencies to pass while blocking higher frequencies. Direct current has a frequency of zero Hz while the ripple voltage of a full-wave rectifier has a frequency of 100 Hz. Connecting the low pass filter will allow the d.c. to pass while blocking the ripple voltage, resulting in a smoother output voltage.

The low pass filter shown in Fig. 5.9 does, however, increase the output resistance which encourages the voltage to fall as the load current increases. This can be reduced if the resistor is replaced by a choke, which has a high impedance to the ripple voltage but a low resistance which reduces the output ripple without increasing the output resistance.

Diode and capacitor ratings

The selection of diodes and capacitors for a rectified a.c. supply needs some careful consideration. It is important to choose a diode which will not be damaged by either the current flowing through it or the voltage developed across it. The capacitor discharges very little during the half cycle that the transformer changes from its maximum positive to maximum negative value and, therefore, the voltage across the diode can be equal to almost twice the maximum value of the a.c. supply. The diode must withstand this maximum reverse voltage without damage, and, therefore, a diode must be chosen which has a maximum reverse voltage rating of about four times the rms voltage of the transformer secondary voltage. This will then take account of the maximum reverse voltage plus any mainsborne interference which may be present.

The capacitor should have a voltage rating at least equal to the peak value of the transformer secondary voltage, so that the bumpy output from the diodes will not break down the capacitor insulation. Also, over much of the a.c. cycle, the load current is supplied by the capacitor, which must be capable of supplying the load without overheating. The 'ripple current rating' of a capacitor is usually quoted in the manufacturers' specification, which for an electrolytic capacitor may range from 200 mA upwards.

When the load current of the simple circuit shown

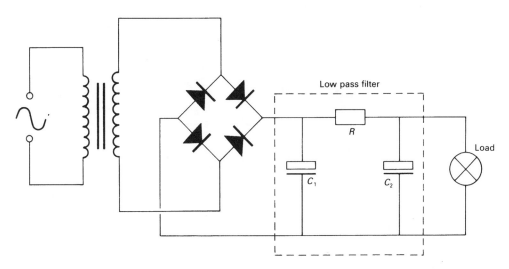

Figure 5.9 Rectified a.c. with low pass filter connected.

in Fig. 5.7 increases, the terminal voltage reduces, mainly because of the resistance of the transformer secondary winding. This reduction in the terminal voltage is called the load regulation or just 'regulation' of the circuit.

REGULATION

Regulation is the percentage change in the output voltage when the load current is increased from zero to full load. A power supply whose output voltage falls rapidly as the load current is increased would be said to have poor regulation. To overcome poor load regulation a number of regulating or stabilising circuits have been developed.

Stabilised power supplies

The power supplies required for electronic circuits must be ripple free, stabilised and have good regulation, that is the voltage must not change in value over the whole load range. A number of stabilising circuits are available which, when connected across the output of the circuit shown in Fig. 5.7 give a constant or stabilised voltage output. These circuits use the characteristics of the zener diode which was discussed in Chapter 4.

ZENER DIODE STABILISER

Figure 5.10 shows an a.c. supply which has been rectified, smoothed and stabilised. In designing the circuit we must give consideration to the following points. The value of the current-limiting resistor must be chosen so that the power rating of the zener diode is not exceeded when operating. In the case of a BZY88-type zener diode, the maximum power rating is 500 mW and, therefore, the maximum current which can flow through a 9.1 V zener diode will be given by

$$I_{max} = \frac{\text{Power W}}{\text{Volts V}} = \frac{500 \times 10^{-3} \text{W}}{9.1 \text{ V}} = 55 \text{ mA}$$

When there is no load connected the output current from the smoothing circuit will flow through the zener diode, and this must be limited to no more than 55 mA by the series resistor, to prevent the diode overheating. However, because the power dissipation of the zener diode is determined for free air, we will base further calculations upon a maximum current of only 40 mA to avoid any possibility of the diode overheating. The output voltage from the smoothing capacitor will be greater than the rms voltage of the transformer secondary because the capacitor will charge up to the maximum value of the secondary voltage. The voltage across the capacitor of Fig. 5.10 will, therefore, be approximately 15 V after deducting the volt drop across the two diodes from the maxi-

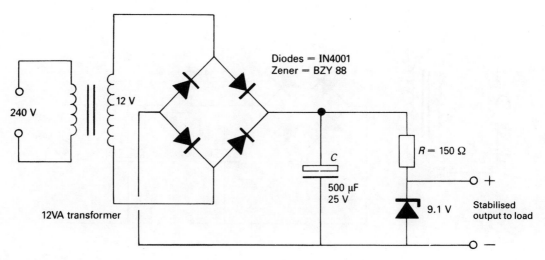

Figure 5.10 Stabilised d.c. supply.

mum value. This voltage will become the input voltage to the stabilising circuit and the voltage levels are shown in Fig. 5.11.

If we assume that 40 mA flows through the series resistor and diode, the resistor must have a value of

$$R = \frac{V}{I} = \frac{6\text{ V}}{40 \times 10^{-3}\text{ A}} = 150\ \Omega$$

The smoothing capacitor must be capable of supplying the ripple volts and have a voltage rating of at least the peak transformer voltage.

The values of all the circuit components are as shown in Fig. 5.10.

When the load current is a maximum, sufficient current must flow through the zener diode to turn it on to the breakdown part of the characteristic shown in Fig. 4.7. This will be at least 5 mA for a small diode and is usually achieved by making the input to the stabilising circuit several volts greater than the output.

The circuit shown in Fig. 5.10 will, therefore, provide a stabilised voltage output of approximately 9 V and be capable of delivering a maximum load current of 35 mA.

The output voltage of this simple stabilised circuit may vary slightly because the slope of the zener diode characteristic is not perfectly linear. To give a constant output, the current through the zener diode must remain constant. Two ways of achieving this are to use a second diode as a stabiliser or a transistor as a constant current source.

Figure 5.12 shows a slightly improved but more expensive circuit which makes use of a second zener diode to compensate for regulation and, therefore, provide a more stable output.

The circuit shown will, however, only supply a load of about 15 mA and a much more elegant solution is to use a transistor as a constant current source. In this case the output current is multiplied by the current gain of the transistor without increasing the demand on the zener diode. In the case of Fig. 5.13 the output will be about one ampere. The smoothing capacitor must be big enough to prevent the ripple voltage dropping below 9.1 V while supplying the load current and, therefore, the parallel capacitors have been used in this circuit to supply the relatively large load. (In electronic circuits one ampere represents a lot of current!)

Fixed voltage series regulators

For really heavy current applications (2 A to 5 A), a voltage regulator connected in series with the load is the most elegant and least expensive solution. This is a three-terminal integrated circuit which incorporates zener diodes, transistors, resistors and capacitors in one package. The internal circuitry provides automatic thermal overload protection and to achieve maximum performance the internal power dissipation must be kept below 50 W by mounting the voltage regulator on a heat sink. Figure 5.14 shows the basic circuit diagram for a fixed voltage regulator with an output of 12 V and capable of supplying output currents up to 5 A. Other output voltages and ratings are available as three-terminal packages of integrated circuits and these are tabulated in Appendix I.

The system or block diagram for a d.c. power supply is shown in Fig. 5.15. You should by now be reasonably familiar with the contents of each section. The transformer converts the a.c. mains supply to a safer, lower value a.c. supply. The rectifier converts the a.c. supply into a bumpy d.c. supply. The filter, usually a high-value electrolytic capacitor, smooths out the bumpy d.c. supply to no more than a ripple. The stabiliser, a circuit incorporating a zener diode, fixes and maintains the voltage level at a precise value. The shape of the waveform as it would appear on an oscilloscope at each stage is also shown in Fig. 5.15.

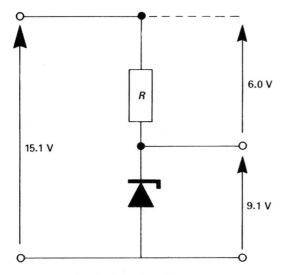

Figure 5.11 Voltage levels for the stabilising circuit given in Figure 5.10.

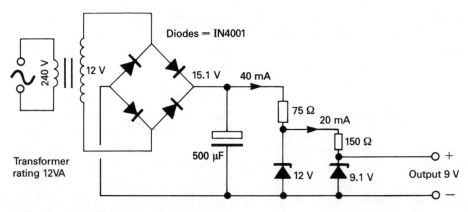

Figure 5.12 Stabilised d.c. supply with compensation for regulation.

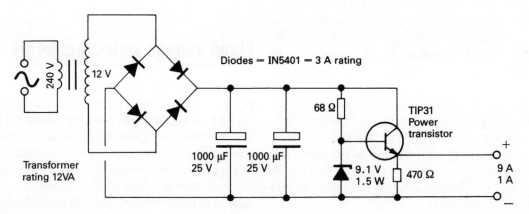

Figure 5.13 Stabilised d.c. power supply.

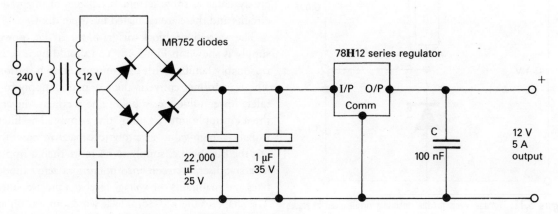

Figure 5.14 Stabilised d.c. power supply using a fixed voltage series regulator.

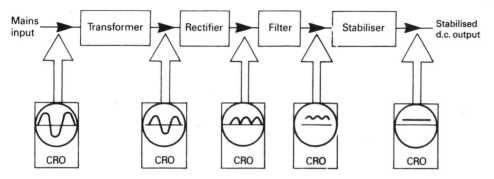

Figure 5.15 A system or block diagram for a d.c. power supply showing the wave shape at each stage.

POWER CONTROL USING THE THYRISTOR

The thyristor is a power rectifier which is triggered into its forward conducting state by applying a small positive pulse or continuous d.c. voltage to the 'gate' connection. It is also called a silicon *controlled* rectifier (SCR) since the gate pulse *controls* the switching of the device which is made from silicon.

Once switched into the forward conducting state by the gate pulse, the device cannot be switched off until the applied voltage falls to zero. Thyristors, therefore, lend themselves to the control of power in a.c. circuits because the voltage waveform of the a.c. mains passes through zero twice on each cycle, which allows the device to be switched off. Two methods are generally employed to control a thyristor in an a.c. circuit, burst trigger control or phase control.

BURST TRIGGER CONTROL

Burst triggering, synchronous triggering or zero voltage triggering are different names for the same method of thyristor control which is directly comparable with traditional methods of control in which power is switched on and off for various intervals of time as shown in Fig. 5.16. In this case the thyristor is triggered only on the first, third and fourth cycles, but more power could be delivered to the load by triggering every cycle, or less power by triggering fewer cycles. Mechanical or thermal inertia is used to smooth out the effects of this bumpy waveform on the load. Switching the thyristor on as the mains waveform passes through zero has the advantage of reducing the effect of mainsborne interference which occurs if the thyristor is switched on part way through a cycle, but a much smoother and more desirable method of controlling the mains waveform is provided by phase control.

PHASE CONTROL

Phase control is a technique used for varying the effective power to a load by rapidly switching the a.c. supply connected to the load for a controlled and adjustable fraction of every cycle. Triggering the thyristor early in the cycle will deliver more power to the load. Triggering the thyristor later in the cycle will reduce the power available. The angle at which the thyristor is turned on is called the trigger angle α (Greek letter alpha) and the angle through which the thyristor conducts is called the conduction angle θ (Greek letter theta). Figure 5.17 shows a triggering angle of about 45°.

This method of control is different to burst triggering because periods of conduction occur during *every* positive half cycle of the mains voltage. The supply to the load is less bumpy and this method of control is, therefore, ideally suited for applications where a smooth supply of controllable power is required. Phase control is essential if filament lamps are to be dimmed electronically, otherwise lamp flicker is troublesome.

Figure 5.18 shows a simple thyristor-controlled circuit which could be used for motor speed control or as a lamp dimmer.

The trigger angle and, therefore, the power available to the load is adjusted by varying the resistor R_2. Resistors R_1 and R_2 act as a voltage divider for the circuit. The diode D_2 ensures that only a positive voltage is applied to the gate of the thyristor. The diode D_1 prevents the negative half cycle of the mains waveform being presented to the gate of the thyristor.

Adjusting the variable resistor R_2 will vary the voltage at the gate of the thyristor. This will change the trigger angle and, therefore, vary the power avail-

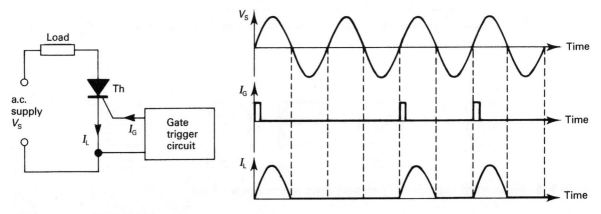

Figure 5.16 Burst trigger control of a thyristor.

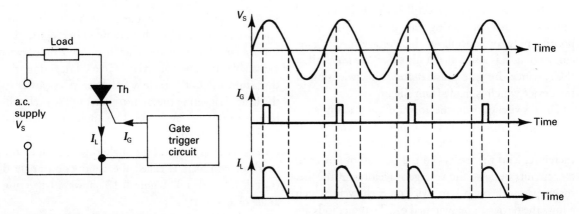

Figure 5.17 Phase control of a thyristor.

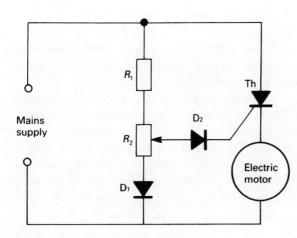

Figure 5.18 A simple thyristor-controlled circuit.

able to the load giving, in the case of Fig. 5.18, speed control of the electric motor.

The switching of motor and discharge lighting circuits causes problems for electronic circuit designers because of the inductive nature of these loads. The inductance generates mainsborne and radio frequency interference which disrupts the triggering of the thyristor. The circuit shown in Fig. 5.19 is essentially the same as Fig. 5.18 except for the addition of a 1 kΩ resistor in the gate circuit. This is known as a '*bleed*' resistor which makes the gate less sensitive to the interference generated by the inductive load and prevents unwanted triggering which allows more accurate speed control of the motor.

Varying the 5 kΩ resistor will vary the trigger angle of the thyristor. The 10 kΩ and 5 kΩ resistors act as a voltage divider which, in this case, causes approximately one third of the voltage available to be dropped across the variable resistor. The diode D_2 will ensure

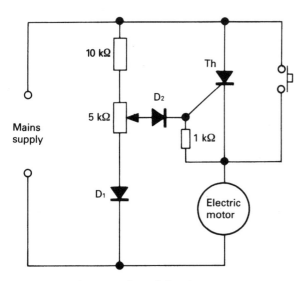

Figure 5.19 Thyristor speed control of an electric motor.

that only a positive voltage is applied to the gate of the thyristor. Diode D_1 prevents the negative half cycle of the mains waveform being presented to the thyristor gate. The speed of the electric motor is controlled by the trigger action of the thyristor gate. However, it cannot run at full speed because the thyristor is a half-wave device, which cuts out the negative half cycle of the mains supply as shown in Fig. 5.17. Closing the switch will cut out the thyristor connecting the mains supply to the motor, which will then run at its maximum speed. Thryistor speed control is only possible in this circuit with the switch open.

Flywheel diode

When highly inductive loads such as motor or discharge lighting circuits are to be switched in electrical installation work the IEE Regulations recommend that the functional switches be rated at *twice* the total steady current of the circuit. This is to avoid excessive wear of the switch contacts resulting from the back emf which is generated when inductive circuits are switched. (See Chapter 10 of *Advanced Electrical Installation Work*.)

A thyristor switches the load on and off during every half cycle and, when used to switch inductive loads, it has been found that the back emf causes the trigger circuit to lose control and, therefore, the current continues to flow in the load when the thyristor has attempted to switch off. This problem can be eliminated by fitting a *bypass* or *flywheel* diode which is connected in parallel with the load. This allows the reverse current generated by the back emf to circulate harmlessly around the loop formed by the load and the diode, leaving the thyristor to be triggered normally by the gate signal.

A flywheel diode is fitted to the motor circuit, shown in Fig. 5.20. The variable resistor R varies the trigger angle and, therefore, the point at which the thyristor switches on during each positive half cycle. This varies the power available to the motor which varies the speed. The thyristor switches off at the end of each half cycle and the motor generates a back emf, which circulates a clockwise current around the parallel circuit made up of the motor and the flywheel diode D. The energy contained in the back emf is then dissipated harmlessly in the resistance of this parallel circuit.

If the motor fails to operate, the following simple checks can be made to the circuit.

1. Test for a satisfactory output voltage from the rectifier at fuses F_1 and F_2.

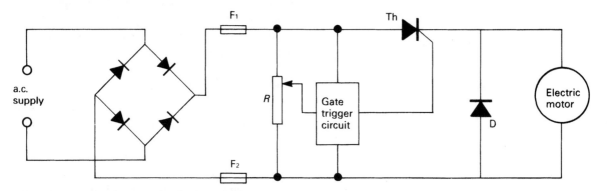

Figure 5.20 Thyristor speed control of an electric motor fitted with a flywheel diode.

2. Connect a short piece of cable between the anode and cathode of the thyristor. This link will short out the thyristor and test the motor. If the motor runs, the fault is in *either* the thyristor *or* the gate trigger circuit. Remove the link.

3. Disconnect the gate trigger circuit connections to the variable resistor and the thyristor gate. Turn the variable resistor R to the middle position and connect a short piece of cable from the wiper connection of the variable resistor to the thyristor gate connection. If the motor operates, the fault is in the trigger unit. If the motor fails to operate, the thyristor is probably faulty and should be removed from the circuit and tested as described in Fig. 4.23.

Three-phase power control using thyristors

A three-phase supply can be used to drive a d.c. motor if the circuit shown in Fig. 5.21 is assembled. The thyristor trigger pulses are electronically separated by 120° so that only one thyristor conducts at any one time.

When TH1 is triggered, current passes to the load and returns via either D2 or D3, whichever diode is the most negative at each instant of time. In practice current will return by D2 for a short period of time and then by D3. When TH2 is triggered 120° later, current passes to the load and returns via either D1 or D3. TH3 is then triggered and the current returns by D1 or D2. In this way, the load is presented with a unidirectional supply which contains only a small a.c. ripple.

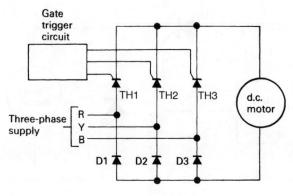

Figure 5.21 Three-phase power control using thyristors.

POWER CONTROL USING TRIAC

The thyristor is a half-wave device, allowing control of only half the available power at any one time, which is uneconomical. The triac is a single three-terminal device which is equivalent to two back-to-back thyristors. This allows it to conduct on both halves of the supply waveform and to be triggered on with either polarity of gate signal. Figure 5.22 shows a simple triac circuit and the resulting waveforms.

DIAC AS A TRIGGER DEVICE

The diac is used mainly as a trigger device for thyristors and triacs. It is a two-terminal device which is equivalent to two back-to-back zener diodes. When a pre-determined voltage is reached, typically 30 V, the diac conducts providing the trigger pulse to the gate of the triac. The trigger pulse occurs on both positive and negative half cycles and so the triac can take advantage of all the power available.

Figure 5.23 shows a diac-triggered triac controlling a lamp dimmer circuit. When the capacitor C has charged up to the diac switching voltage, say 30 V, the diac will conduct, the capacitor will discharge through the diac, providing a trigger pulse to the triac, which will switch on and control the current flowing through the load and, therefore, the power to the lamp. Increasing the value of resistance R will increase the time taken for the capacitor to reach the diac switching voltage and, therefore, the point of switch on will be delayed. (See Chapter 8, Charging capacitors) Reducing the value of resistance will reduce the time constant of the C-R circuit and the triac will switch on earlier in each half cycle, delivering more power to the lamp which will illuminate brightly. The waveforms will be similar to those shown in Fig. 5.22 with the variable resistor controlling the trigger angle of the triac and, therefore, the lamp brightness. The R-C circuit on the right of the circuit diagram and connected across the triac is known as a *snubber circuit*.

SNUBBER NETWORK

Triacs are also susceptible to false triggering caused by mainsborne interference spikes. The *snubber circuit* takes out these instantaneous changes and prevents the triac triggering inadvertently. When the triac is used with inductive loads, it is essential to connect a *snubber network* across the triac. Typical values are 100 Ω and 0.1 μF as shown in Fig. 5.23.

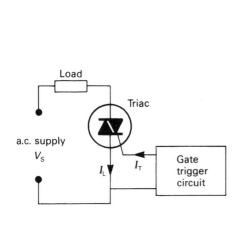
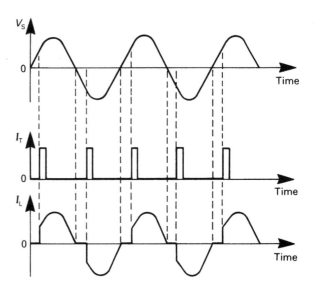

Figure 5.22 Simple triac circuit and waveforms.

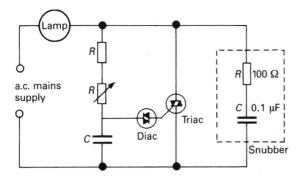

Figure 5.23 Lamp dimmer circuit using diac triggering.

Manufacturers' data sheet information and practical information on thyristors and triacs is included in Appendix J.

Amplifiers

Amplification is one of the most important functions in electronics. Amplifiers are usually concerned with converting small changes of voltage at the input of the amplifier into larger changes of voltage at the output. They are used in hi-fi equipment to increase the signal strength from a tape, record or compact disc before it can be usefully played through the speaker system. The very small signal from a transducer such as a strain gauge can only be used for display purposes after being amplified as described in Chapter 12.

The amount by which an amplifier amplifies is known as the *gain* of the amplifier and has the general symbol A:

$$\text{gain} = A = \frac{\text{output signal}}{\text{input signal}}$$

In a voltage amplifier the amplified quantity is voltage and, therefore, the gain of the amplifier can be determined by

$$\text{voltage gain} = A_V = \frac{\text{output voltage}}{\text{input voltage}}$$

In a similar way

$$\text{current gain} = A_I = \frac{\text{output current}}{\text{input current}}$$

The feature of an amplifier which makes it important is that it can provide power gain and not just voltage gain. If we multiply the voltage gain by the current gain we obtain the power gain:

$$\text{power gain} = A_P = \frac{\text{output power}}{\text{input power}}$$
$$= A_V \times A_I$$

In electronics, as in life, we never get something for nothing and this power gain must come from another energy source. In practice, every electronic amplifier must be supplied with a constant voltage from either a battery or a d.c. stabilised power supply circuit such as those described earlier in this chapter.

In many practical circuits the required voltage gain cannot be provided by any one amplifier stage. In this case the separate amplifier stages are connected together in series or cascade, output to input, and the total voltage gain can then be found by multiplying together the individual voltage gains. The total voltage gain of a number of stages is given by

$$\text{total gain } A_V = A_{V_1} \times A_{V_2} \times A_{V_3} \text{ etc.}$$

OPERATIONAL AMPLIFIER

The actual amplifier circuit can be made up from individual discrete transistors, or the most complex arrangement of transistors, resistors and capacitors can be integrated on to a tiny silicon chip which is then known as an integrated circuit or IC. When an amplifier is made in this way it is called an operational amplifier or op amp. The circuit symbol and appearance are shown in Fig. 5.24. This is the easy way to amplify weak signals and although many different types of op amp are available, the 741 is the industry standard. The gain of a 741 op amp is about 100 000 compared with approximately 100 for a transistor amplifier. For most practical purposes a gain of this magnitude is much too large and unstable but the actual performance can be modified with external components. This is found to reduce the distortion of the output signal, increase the range of frequencies which can be amplified and reduce but stabilise the gain of the amplifier. These advantages outweigh the loss of gain which can easily be increased by using two or more op amp stages.

The basic circuit diagram for an op amp voltage amplifier is shown in Fig. 5.25. The output voltage of this op amp will be greater than the input voltage by a factor determined by gain A_V and in antiphase to it, because the input is connected to the inverting (–)

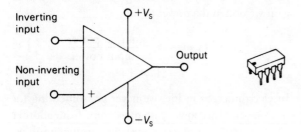

Figure 5.24 Circuit symbol and appearance of a 741 op amp.

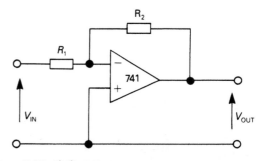

Figure 5.25 The basic op amp circuit.

terminal and negative feedback is provided by R_2. The gain is controlled by the ratio of R_2/R_1 and in the case of negative feedback, we add a negative sign to the formula to obtain the equation

$$A_V = \frac{-R_2}{R_1}$$

Therefore, the output V_{OUT} of Fig. 5.25 will be

$$V_{OUT} = -V_{IN} \frac{R_2}{R_1} \text{ volts}$$

If the values of R_2 and R_1 in Fig. 5.25 were 200 kΩ and 1 kΩ, the gain of the op amp would be 200. With an input signal V_{IN} of 1 µV, the output would be

$$V_{OUT} = -1 \times 10^{-6} \times \frac{200 \text{ k}\Omega}{1 \text{ k}\Omega} = -0.2 \text{ mV}$$

That is, the output is 200 times greater than the input and inverted. You may feel that 0.2 mV still seems an insignificant voltage but a voltage gain of 200 is significant and these values are what we could expect from a strain gauge circuit or the pick-up of a compact disc player.

When an amplifier is used to amplify the input voltage or current in such a way that the output is an enlarged copy of the input and is not distorted, it is said to be a small signal amplifier. When an amplifier is used to amplify the power of an input signal it is said to be a power amplifier. Figure 5.26 shows the circuit diagram of an audio frequency amplifier. The left-hand side of the circuit, the op amp, is a small signal voltage amplifier which is used to amplify a small signal from, for example, the ear piece jack plug of a tape recorder. The right-hand side of the circuit is the power amplifier which is required to drive the speaker. This is made from a pair of complementary power transistors, one is an n-p-n and the other a

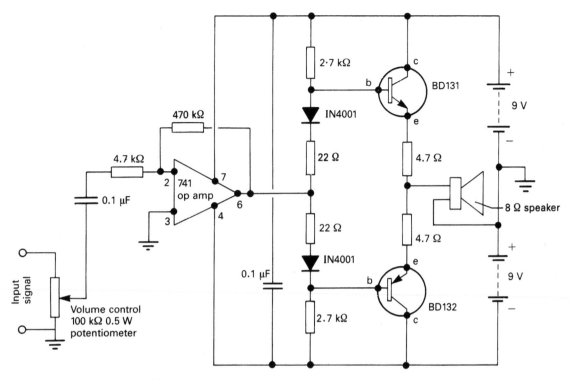

Figure 5.26 An audio frequency amplifier.

p-n-p transistor which have been *matched* so that they have the same gain and other properties. When the voltage on the top transistor is positive the voltage on the bottom transistor is negative and vice versa. The amplification of each half of the voltage waveform is, therefore, shared between the two transistors. A circuit which is constructed in this way is known as a push-pull amplifier. The additional power required to drive the speaker in this circuit comes from the 9 V batteries.

BANDWIDTH OF AN AMPLIFIER

The bandwidth of an amplifier is the range of frequencies within which the gain does not fall below about 0.7 of the maximum gain. The points at which this happens, f_1 and f_2 in Fig. 5.27, are called the 3 dB points. The decibel (dB) scale is used in electronics to compare signal power levels or the loudness of a sound. The unit is named in honour of Alexander Graham Bell (1847–1922), the pioneer of the telephone.

Most audio amplifiers, such as the one shown in Fig. 5.26 are designed to amplify the range of frequencies to which the human ear will respond.

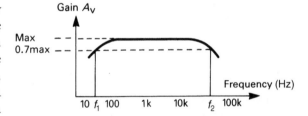

Figure 5.27 Amplifier gain/frequency response curve.

This is approximately 15 Hz to 15 kHz although some manufacturers of good quality audio amplifiers design their circuits with an upper frequency limit of about 40 kHz. Amplifying the frequencies well above the range of human hearing can improve the sound quality within the audio frequency range.

The relationship between gain and the frequency range of a 741 op amp is shown in Fig. 5.28. Reducing the gain increases the frequency range. If the circuit was designed to have a gain of say 1000, the amplifier will respond to frequencies in the range 1 Hz to 1 kHz. However, this upper frequency limit is too low if the circuit is to be used as an audio amplifier. The gain will, therefore, need to be reduced to between 10 and 15 to obtain an upper frequency

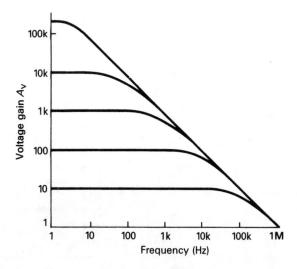

Figure 5.28 Frequency response of a 741 OpAmp for various values of gain.

limit of about 20 kHz if the circuit is to be used as an audio amplifier.

Filters

Filters are circuits which allow some a.c. frequencies to pass through them more easily than others. There are many instances in electronics where the frequency bandwidth of a circuit must be restricted, for example, to reduce the signal strength-to-noise ratio which is directly proportional to bandwidth.

If an old 78 rpm gramophone record is played on a modern wide frequency range hi-fi system, it may be difficult to hear the music above the surface noise. However, the same record played on a narrow frequency range 1950 vintage gramophone may sound quite acceptable. This is because the bandwidth is wide enough to pass the music but narrow enough to limit the noise frequencies. Two of the most simple filters are called *low pass* and *high pass* filters.

Figure 5.29 shows a simple first-order high pass filter which is made up of a capacitor and resistor. The reactance (resistance to a.c.) of a capacitor reduces as the frequency increases and, therefore, at low frequencies the reactance will be high and at high frequencies the reactance will be low. This circuit will, therefore, allow high frequencies to pass more easily than low frequencies and hence, is called a high pass filter. High pass filters can be used to block d.c. when a number of a.c. amplifier stages are connected together.

Figure 5.30 shows a simple first-order low pass filter. The resistor and capacitor are connected the opposite way round and the circuit will, therefore, pass low frequencies better than high frequencies, hence it is called a low pass filter. Low pass filters can be used to block a.c. ripple on d.c. power supplies as shown in Fig. 5.9 which shows a slightly improved low pass filter arranged in a π configuration.

Testing audio amplifiers

If the audio amplifier circuit shown in Fig. 5.26 is assembled on a prototype board as shown in Fig. 3.21, a number of tests can be carried out.

The circuit must be built *exactly* as shown in the circuit diagram but a 10 Ω 1 W resistor is used as the load in place of the 8 Ω speaker. The component pin

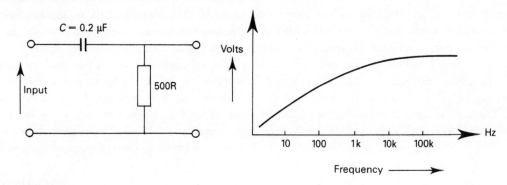

Figure 5.29 High pass filter.

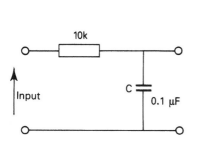
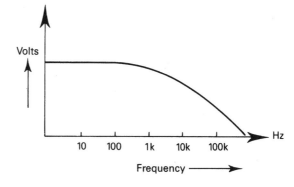

Figure 5.30 Low pass filter.

connection information given in Fig. 5.31 will help with the actual construction.

Adjust the settings on a double output power supply unit (PSU) to 9 V and connect with flying leads to the circuit assembled on the prototype board, making sure that the polarities are as shown in the circuit diagram.

Connect a signal generator to the amplifier input terminals and adjust it to give a sine wave at 1 kHz which is the industrial standard test frequency for audio amplifiers. Adjust the volume control to the maximum position. Set all the CRO controls to the calibrate position and connect the leads across the amplifier output terminals, that is across the 10 Ω load resistor. You are now ready to carry out the tests.

TESTING FOR SIGNAL DISTORTION

Increase the voltage output from the signal generator while observing the amplifier output waveform on the CRO. You should observe a sine wave increase in magnitude until it begins to distort at the top and bottom of the waveform. That is, the sine wave will *clip* or flatten at the top and bottom of the waveform. Reduce the input until the output waveform is once more sinusoidal and measure the input voltage being delivered by the signal generator. This is the maximum value of input voltage which can be applied to the amplifier before distortion occurs.

TESTING FOR GAIN

Adjust the voltage output from the signal generator to a value of about 20 mV peak to peak at a frequency of 1 kHz. Measure the output and input voltage with the CRO and calculate the amplifier gain by dividing the output voltage value by the input voltage value.

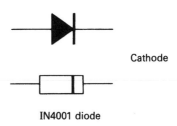

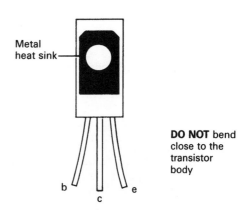

Pin identification of both transistors

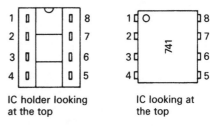

Pin identification of 741 IC

Figure 5.31 Component pin connection information to be used when building the audio amplifier circuit shown in Figure 5.26.

MEASURING THE BANDWIDTH

To be suitable as an audio amplifier the circuit shown in Fig. 5.26 must amplify most of the frequencies to which the human ear will respond, that is between about 15 Hz and 15 kHz. To measure the actual bandwidth of the amplifier we must, therefore, make a series of gain measurements over this range of frequencies in order to plot a gain/frequency response curve such as that shown in Fig. 5.27. To do this we must vary the output frequency while maintaining a constant output voltage from a test signal source such as a signal generator as follows.

Set the output frequency of the signal generator to 10 Hz and adjust the voltage level to 20 mV peak to peak. Use the CRO to measure the voltage at the input and output terminals of the amplifier. The gain at this frequency can then be calculated as before and the results entered in Table 5.1. Next, set the output frequency of the signal generator to 30 Hz and adjust the voltage to 20 mV peak to peak. Use the CRO to measure the voltage at the input and output terminals of the amplifier and once more calculate the gain and enter the results in Table 5.1. Continue to take and enter readings in this way until Table 5.1 is completed.

The gain/frequency response curve can then be plotted on the Log/Lin graph paper as shown in Fig. 5.32 using the gain and frequency values tabulated in Table 5.1. The logarithmic horizontal scale is necessary to accommodate the large frequency range of an audio amplifier. When all the points have been plotted they can be joined to form a bandwidth curve which should be similar to that shown in Fig. 5.27. The bandwidth is the range of frequencies enclosed by the upper and lower frequency points when the gain is equal to 0.7 of the maximum gain. What is the bandwidth of your amplifier? The bandwidth of an expensive music centre amplifier might be between 20 Hz and 40 kHz while the telephone

Table 5.1 A table of results which is suitable for measuring the bandwidth of an audio amplifier

Frequency (Hz)	V_{IN}	V_{OUT}	Gain = A_v = V_{OUT}/V_{IN}
10	20 mV		
30	20 mV		
60	20 mV		
100	20 mV		
500	20 mV		
1000	20 mV		
4000	20 mV		
6000	20 mV		
8000	20 mV		
10 000	20 mV		
20 000	20 mV		

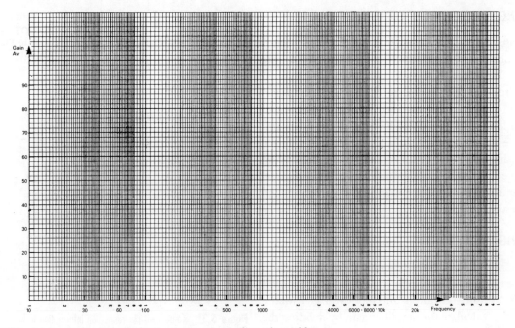

Figure 5.32 Graph paper for plotting the frequency response curve of an audio amplifier.

system is found to give acceptable voice reproduction with a bandwidth from 300 Hz to 3.4 kHz. Do you think that your amplifier will give acceptable results? You can test it by playing a tape recorder through the amplifier. To do this, remove the 10 Ω load resistor used for testing and connect an 8 Ω speaker across the load. Connect the 9 V supplies from the double PSU as described earlier and connect to the input signal connections of the amplifier, a connection from the ear phones jack plug of a tape recorder. The ear phones signal can in this way be played through the amplifier which you have constructed.

Waveforms

The electricity generating stations generate a sinusoidal alternating voltage waveform from an alternator which is usually driven by a steam turbine. A sinusoidal waveform is a *repetitive, analogue* waveform; repetitive because in the case of the a.c. mains it repeats itself fifty times each second, and analogue because it varies smoothly and continuously between two extremes. In everyday life, analogue systems are all around us. Television, radio and telephone signals are all analogue, the amplifier considered earlier is an analogue amplifier. A car speedometer and fuel gauge are analogue instruments. Analogue electronics is one of *two* main branches of electronics, the other is *digital electronics*. Digital signals do not change smoothly and continuously between voltage levels but have two quite definite levels, either on or off. Digital signals are in the form of electrical pulses whose outputs involve only two levels of voltage, called high or low, where high might be +5 V and low 0 V. The term *mark-to-space* ratio is used in connection with digital signals, particularly square, rectangular and pulse waveforms, and is given by

$$\text{mark-to-space ratio} = \frac{\text{mark time}}{\text{space time}}$$

For a square wave the mark-to-space ratio will be 1 because the mark time and space times are equal. Digital circuits are used in pocket calculators, electronic watches, domestic appliances and motor car control systems and increasingly in communication systems.

In electrical installation work we are mostly concerned with sinusoidal voltage waveforms but in electronics other wave shapes are of great importance. For example, square waves are used for timing and oscillator circuits, sawtooth waveforms for the CRO time base and pulses to trigger a thyristor or triac. Figure 5.33 shows some common waveforms.

Harmonics

All the waveforms shown in Fig. 5.33 can be made up of sinusoidal waveforms having a *fundamental* frequency or reference frequency and *harmonics*. Harmonics are sine waves with frequencies that are multiples of the fundamental frequency. A second harmonic has a frequency which is twice the fundamental frequency and a third harmonic three times the fundamental frequency.

Combining a fundamental sine wave with its many harmonics produces a waveform of a different shape. A triangular wave consists of a fundamental sine wave with a large number of *even* harmonics, two, four, six, eight, etc. A square wave consists of the fundamental plus a large number of *odd* harmonics, three, five, seven, nine etc. Figure 5.34 shows an approximately square wave formed from a fundamental and the third, fifth and seventh harmonics.

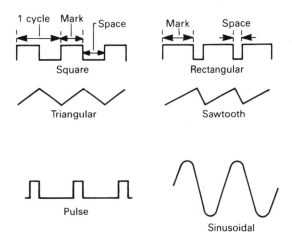

Figure 5.33 Some common waveforms.

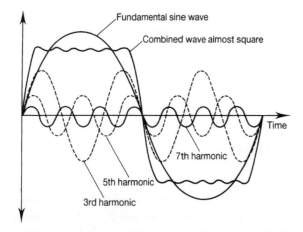

Figure 5.34 Square waves formed from fundamental plus *odd* harmonics.

Signal modulation

It is very difficult to transmit an audio frequency electromagnetic wave such as that produced by a microphone. However, it is relatively easy to transmit a radio frequency wave above about 100 kHz. In radio communications we therefore modify or *modulate* a radio frequency wave so that it 'carries' the shape and, therefore, the intelligence of the audio frequency signal.

In *amplitude modulation* or AM the carrier *amplitude* is modulated by the intelligence signal as shown in Fig. 5.35. The microphone converts sound pressure waves into low frequency electrical signals. An oscillator generates a radio frequency carrier wave and these two signals are fed to the modulator which imprints the microphone signal on to the carrier wave for transmission as shown.

In *frequency modulation* or FM it is the carrier *frequency* which is modulated by the intelligence signal, the carrier amplitude remaining constant. As the modulating signal increases, the carrier frequency increases; as the modulating frequency goes negative, the carrier frequency decreases. Figure 5.36 shows this effect and compares frequency modulation with other types of modulated signal.

In *pulse modulation* a very high frequency carrier is transmitted in pulses rather like very rapid Morse code. Pulse modulation is used mainly for RADAR and very high frequency communication systems.

The pulses are modulated in one of three ways, pulse amplitude modulation (PAM), pulse position modulation (PPM) or pulse width modulation (PWM).

PAM modifies the amplitude of the pulses so that the transmitted signal looks very similar to an AM signal. PPM changes the phase position of the pulses relative to the normal in order to transmit the information.

PWM uses the modulating signal to vary the width of the carrier pulse, a wide pulse means a high positive value and a narrow pulse a high negative value. Figure 5.36 compares pulse width modulation with AM and FM.

Following transmission of the AM, FM or PWM signals the receiving equipment must once more separate the intelligent information from the carrier wave, a process known as demodulation and Figs. 9.11 and 9.13 look at these systems.

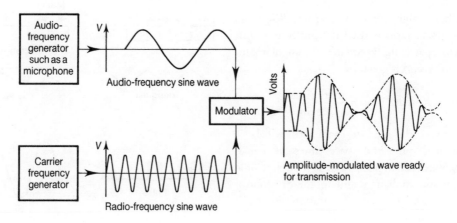

Figure 5.35 Amplitude modulated signal.

ELECTRONIC CIRCUITS IN ACTION

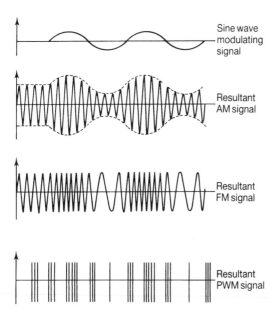

Figure 5.36 Comparison of AM, FM and PWM signals.

Sawtooth waveform generator

If a circuit is constructed as shown in Fig. 5.37 the capacitor will charge up at a rate which is determined by the values of *C* and *R* when the supply is switched on. (See Chapter 8 under Charging capacitors for the relevant theory.) However, if a neon lamp or a thyristor is connected across the capacitor so that the capacitor is discharged quickly at some predetermined point during the lower linear part of the capacitor growth curve, then a sawtooth waveform is generated which has a relatively long rise time and a short discharge time. When the capacitor is discharged to zero volts the neon lamp or thyristor will switch off and the capacitor will once more begin to charge up to the predetermined discharge point and the sawtooth waveform will, therefore, be generated continuously. The long rise time to short discharge time is characteristic of a sawtooth waveform. A sawtooth waveform generator is used to drive the electron beam across the X axis of a CRO screen. During the long rise time the beam sweeps from the left- to the right-hand side of the screen. Flyback occurs during the short discharge period placing the beam once more on the left-hand side ready for a further sweep across the screen.

Transistor switching using a capacitor

This is another application of the circuit theory associated with *charging capacitors* discussed in Chapter 8. If a circuit is constructed as shown in Fig. 5.38 the lamp will illuminate when the supply is switched on because a signal will be applied to the base region of the transistor, but as the capacitor charges up, the voltage across the resistor will fall to

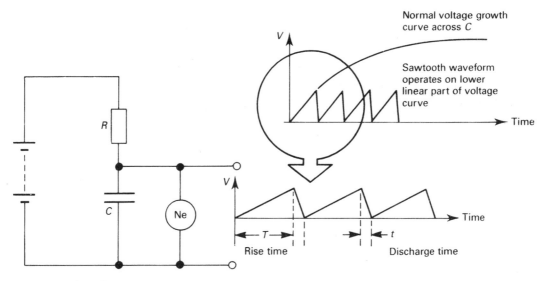

Figure 5.37 Sawtooth waveform generator circuit.

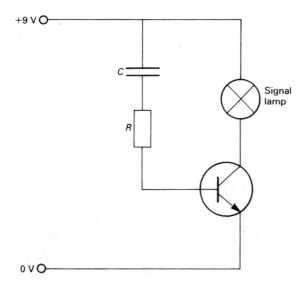

Figure 5.38 Transistor switching using a capacitor.

zero, switching off the signal to the transistor base, which will switch off the transistor and the signal lamp. The capacitor charging current can be used to switch the transistor on or off at a rate which is dependent upon the value of the C-R time constant. This principle can be applied to the electronic circuits known as multivibrators.

MULTIVIBRATORS

Multivibrators are a class of electronic switching circuits which are also known as *relaxation oscillators* because their operation is characterised by a period of vigorous activity followed by a period of relaxation. They produce an output which is generally a series of square or rectangular pulses by switching between two transistors. Figure 5.39 shows the circuit diagram for an *astable* multivibrator. Astable means not stable or free running and therefore, the circuit generates a continuing train of pulses at the output. The two transistors are coupled together, so that while one transistor is switched on, the other is switched off. Each transistor then switches automatically to its other state and, as a result, the output voltage which can be taken from the collector of either transistor is alternatively *high* (+8 V) or *low* (almost 0 V) generating a series of theoretically square pulses. In practice, the switch over from one transistor to another is not instantaneous and, therefore, the output will not be perfectly square as shown by Fig. 5.40. The length of time each transistor is switched on depends upon how long it takes the capacitors to charge up through the resistors. TR1 is switched by R_2 C_2 and TR2 by R_1 C_1. If $C_1 = C_2$ and $R_1 = R_2$ the 'off' and 'on' times for each transistor will be equal and a square wave will be generated. The *rate* at which the signal lamps will flash can be varied by changing the capacitor and resistor values as suggested by Table 5.2. Reducing the values of C and R will reduce the circuit time constant and increase the lamp flashing rate. Using different values of C_1 and C_2 or different values of R_1 and R_2 will vary the mark-to-space ratio and a rectangular waveform will be generated which will result in an uneven flashing rate of the two signal lamps.

It may help your understanding of the circuit if you were to assemble the multivibrator circuit on an S-DeC; vary the component values as suggested by Table 5.2 and observe the results.

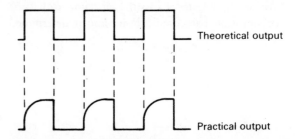

Figure 5.40 Astable multivibrator output.

Table 5.2 A results table for the multivibrator circuit investigation

C_1, C_2	220 µF	220 µF	220 µF	22 µF
R_1, R_2	4.7 kΩ	10 kΩ	22 kΩ	10 kΩ
Number of flashes per minute				
Time constant				

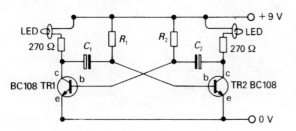

Figure 5.39 Astable multivibrator.

6

TESTING ELECTRONIC CIRCUITS

The use of electronic circuits in all types of electrical equipment has increased considerably over recent years. Electronic circuits and components can now be found in leisure goods, domestic appliances, motor starting and control circuits, discharge lighting, emergency lighting, alarm circuits and special effects lighting systems. There is, therefore, a need for the installation electrician and service engineer to become familiar with some basic electronic test equipment, which is the aim of this chapter.

Test instruments

Electrical installation circuits usually carry in excess of one ampere and often carry hundreds of amps. Electronic circuits operate in the milliampere or even micro-ampere range. The test instruments used on electronic circuits must have a *high impedance* so that they do not damage the circuit when connected to take readings. All instruments cause some disturbance when connected into a circuit because they consume some power in order to provide the torque required to move the pointer. In power applications these small disturbances seldom give rise to obvious errors, but in electronic circuits, a small disturbance can completely invalidate any readings taken. We must, therefore, choose our electronic test equipment with great care. Let us consider some of the problems.

Let me first of all define what is meant by the terms error and accuracy used in this chapter. When the term *error* is used it means the *deviation of the meter reading from the true value* and *accuracy* means the *closeness of the meter reading to the true value*.

INSTRUMENT ERRORS

Consider a voltmeter of resistance 100 kΩ connected across the circuit shown in Fig. 6.1 (a).

Connection of the meter loads the circuit by effectively connecting a 100 kΩ resistor in parallel with the circuit resistor as shown in Fig. 6.1 (b) which changes the circuit to that shown in Fig. 6.1 (c).

Common sense tells us that the voltage across each resistor will be 100 V but the meter would read about 66 V because connection of the meter has changed the circuit. This loading effect can be reduced by choosing instruments which have a very high impedance. Such an instrument imposes less load on the circuit and gives an indication much closer to the true value.

The deflection torque of most instruments is proportional to current and since current $I = V/Z$ and $Z^2 = R^2 + X_L^2$ and $X_L = 2\pi f L$ the instrument is also frequency dependent. The important practical consideration is the *frequency range* of the test instrument. This is the range of frequencies over which the instrument may be considered free from frequency errors and is indicated on the back of the instrument or in the manufacturer's information. Frequency limitations are not a normal consideration for an electrician since electrical installations operate at the fixed mains frequency of 50 Hz.

The scale calibration of an instrument assumes a sinusoidal supply unless otherwise stated. Non-sinusoidal or complex waveforms contain harmonic frequencies which may be outside the instrument frequency range. The chosen instrument must, therefore, be suitable for the test circuit waveform.

The maximum permissible errors for various instruments and their applications are indicated in

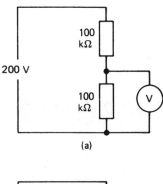

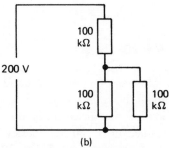

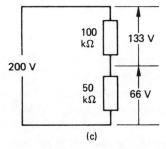

Figure 6.1 Circuit disturbances caused by the connection of a voltmeter.

British Standard 89. When choosing an instrument for electronic testing an electrician or service engineer will probably be looking for an instrument with about a 2% maximum error, that is 98% accurate. Instrument manufacturers will provide detailed information for their products.

Errors are not only restricted to the instrument being used, operators can cause errors too.

OPERATOR ERRORS

These are errors such as mis-reading the scale, reading 28.3 and tabulating 23.8 or reading the wrong scale on a multi-range instrument. The test instrument must be used on the most appropriate scale, don't try to read 12 V on a 250 V scale, the reading will be much more accurate if the 25 V scale is used to read a value of about 12 V.

The type of instrument to be purchased for general use is a difficult choice because there are so many different types on the market and every manufacturer's representative is convinced that his company's product is the best. However, most instruments can be broadly grouped under two general headings: those having *analogue* or *digital* displays.

Analogue and digital displays

ANALOGUE METERS

These meters have a pointer moving across a calibrated scale. They are the only choice when a general trend or variation in value is to be observed. Hi-fi equipment often uses analogue displays to indicate how power levels vary with time, which is more informative than a specific value. Red or danger zones can be indicated on industrial instruments. The fuel gauge on a motor car often indicates full, half full or danger on an analogue display which is much more informative than an indication of the exact number of litres of petrol remaining in the tank.

These meters are only accurate when used in the calibrated position – usually horizontally.

Most meters using an analogue scale incorporate a mirror to eliminate parallax error. The user must look straight at the pointer on the scale when taking readings and the correct position is indicated when the pointer image in the mirror is hidden behind the actual pointer. A good-quality analogue multimeter suitable for electronic testing is shown in Fig. 6.4.

The input impedance of this type of instrument is typically 1000 Ω per volt or 20 000 Ω per volt. depending upon the scale chosen.

DIGITAL METERS

These provide the same functions as analogue meters but they display the indicated value using a seven segment LED (see Fig. 4.11) to give a numerical value of the measurement. Modern digital meters use semiconductor technology to give the instrument a very high input impedance, typically about 10 MΩ and, therefore, they are ideal for testing most electronic circuits.

Figure 6.2 Digital multimeter suitable for testing electronic circuits.

The choice between a meter having an analogue or digital display is a difficult one and must be dictated by specific circumstances. However, if you are an electrician or service engineer intending to purchase a new instrument which would be suitable for electronic testing. I think on balance that a good-quality digital multimeter such as that shown in Fig. 6.2 would be best. Having no moving parts, digital meters tend to be more rugged and, having a very high input impedance, they are ideally suited to testing electronic circuits.

THE MULTIMETER

Multimeters are designed to measure voltage, current or resistance. Before taking measurements the appropriate volt, amp or ohm scale should be selected. To avoid damaging the instrument it is good practice first to switch to the highest value on a particular scale range. For example, if the 10 A scale is first selected and a reading of 2.5 A is displayed, we then know that a more appropriate scale would be the 3 A or 5 A range. This will give a more accurate reading which might be, say, 2.49 A. When the multimeter is used as an ammeter to measure current it must be connected in series with the test circuit as shown in Fig. 6.3(a) When used as a voltmeter the multimeter must be connected in parallel with the component as shown in Fig. 6.3(b).

OHM METER

When using a commercial multi-range meter as an ohm meter for testing electronic components, care must be exercised in identifying the positive terminal. The red terminal of the meter, identifying the positive input for testing voltage and current, usually becomes the negative terminal when used as an ohm meter because of the way the internal battery is connected to the meter movement. To reduce confusion when using a multi-range meter as an ohm meter it is advisable to connect the red lead to the black terminal and the black lead to the red terminal so that the red lead indicates positive and the black lead negative as shown in Fig. 6.4. The ohm meter can then be

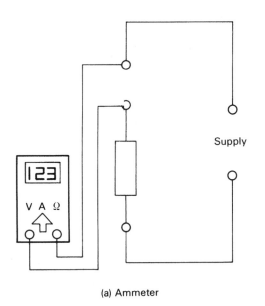

(a) Ammeter

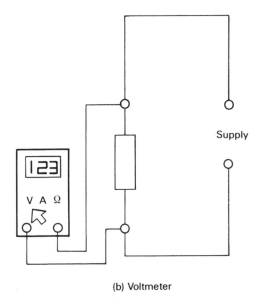

(b) Voltmeter

Figure 6.3 Using a multimeter (a) as an ammeter and (b) as a voltmeter.

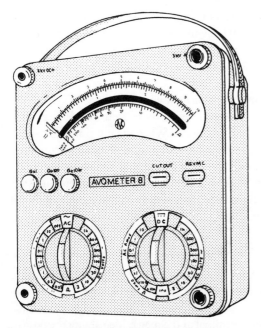

Commonly used multirange instrument

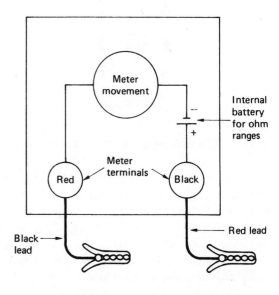

Figure 6.4 Multirange meter used as an ohm meter.

successfully used to test diodes, transistors and thyristors as described in Chapter 4, and resistors and capacitors as described in Chapter 2.

Commercial multi-range instruments reading volts, amps and ohms are usually the most convenient test instrument for an electrician or service engineer, although a cathode ray oscilloscope (CRO) can be invaluable for bench work.

The cathode ray oscilloscope (CRO)

The CRO is probably one of the most familar and useful instruments to be found in an electronic repair service workshop or college laboratory. It is a most useful instrument for two reasons: it is a high impedance voltmeter and, therefore, takes very little current from the test circuit and, secondly, it allows us to 'look into' a circuit and 'see' the waveforms present. The cathode ray is the name given to a high-speed beam of electrons generated in the cathode ray tube and was first used during the Second World War as part of the *Radar* system. The beam of electrons is deflected horizontally across the screen at a constant rate by the *time-base circuit* and vertically by the test voltage. The many controls on the front of the CRO are designed so that the operator can stabilise and control these signals. Figure 6.5 shows the front panel of a simple CRO. Electricians and service engineers who are unfamiliar with the CRO should not be baffled by the formidable array of knobs and switches – take them one at a time – and give yourself time to become familiar with these controls.

The single most important component in the CRO is the cathode ray tube.

CATHODE RAY TUBE

Figure 6.6 shows a simplified diagram of the cathode ray tube. This is an evacuated glass tube containing the *electron gun* components on the left and the fluorescent screen, which the operator looks at, on the right. On the far left of the diagram is the wire filament through which a current is passed. This heats a metal plate called a cathode, which emits the electrons to be accelerated. The *rate* at which the electrons are accepted for acceleration could be modified by making changes to the temperature of the cathode, but in practice it is more convenient to have a metal control grid with a hole in it. By varying the voltage of the control grid it is possible to influence

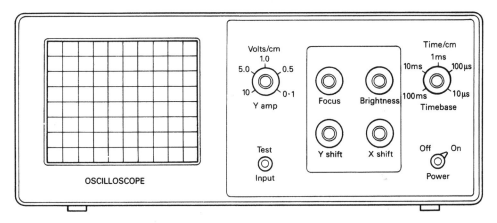

Figure 6.5 Front panel of a simple CRO.

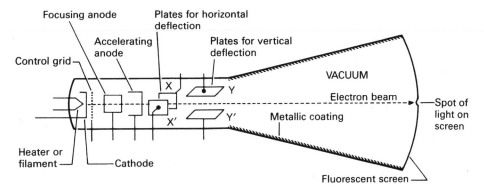

Figure 6.6 Simplified diagram of the cathode ray tube.

the number of electrons passing through the hole in the grid. The electrons which pass through the grid tend to be moving in various directions and the purpose of the next component therefore is to focus the beam. The electrons are then further accelerated by the accelerating anode to give them sufficient final velocity to produce a bright spot on the screen.

The electrons, on emerging through a hole in this anode, pass through two pairs of parallel plates X X′ and Y Y′, each pair being at right-angles to the other. If an electric field is established between X and X′ the beam can be deviated horizontally, the direction and magnitude of the deflection depending upon the polarity of the plates. The negative beam of electrons is attracted towards the more positive plate. Likewise, an electric field between plates Y and Y′ produces a vertical deviation. Therefore, a suitable combination of electric fields across X X′ and Y Y′ directs the beam to any desired point on the screen.

Upon reaching the screen, the electrons bombard the fluorescent coating on the inside of the screen and emit visible light. The brightness of the spot depends upon the speed of the electrons and the number of electrons arriving at that point.

USE OF THE CRO

The function of the various controls is outlined as follows:

1 Power on switch – Switch on and wait a few seconds for the instrument to warm up. An LED usually indicates a satisfactory main supply.
2 Brightness or intensity – This controls the brightness of the trace. This should be adjusted until bright, but not over-brilliant, otherwise the fluorescent powder may be damaged.
3 Scale illumination – This illuminates and highlights the 1 cm square grid lines on the screen.

4 Focus – The spot or trace should be adjusted for a sharp image.
5 Gain controls – Adjust for calibrate.
6 X-shift – The spot or trace can be moved to the left or right and should be centralised.
7 Y-shift – The spot or trace can be moved up or down.
8 TRIG control – This allows the time base to be synchronised to the applied signal to enable a steady trace to be obtained. Set the switch to either Auto or to the Y input which is connected to the test voltage.
9 AC/GND/DC – It is quite common for a signal to be made up of a mixture of a.c. and d.c. Select DC for all signals and AC to block out the d.c. component of a.c. signals. The GND position disconnects the signal from the Y amplifier and connects the Y plates to ground or earth.
10 Chop/Alt – When a double beam oscilloscope is used, it is common practice to obtain the two X traces from one beam by either sweeping the electron beams alternately or by sweeping a very small segment of each beam as the trace moves across the screen, leaving each trace chopped up. Use *chop* for slow time base ranges and *Alt* for fast time base ranges.
11 Connect the test voltage to the CRO leads and adjust the calibrated Y-shift (volts/cm) and time base (time/cm) controls until a steady trace fills the screen.

USE OF THE CRO TO MEASURE VOLTAGE AND FREQUENCY

The calibrated Y-shift, time base and 1 cm grating on the tube front provide us with a method of measuring the displayed waveform.

With the test voltage connected to the Y-input, adjust all controls to the calibrate position. Adjust the X and Y tuning controls until a steady trace is obtained on the CRO screen, such as that shown in Fig. 6.7.

To measure the voltage of the signal shown in Fig. 6.7 count the number of centimetres from one peak of the waveform to the next using the centimetre grating. This distance is shown as 4 cm in Fig. 6.7. This value is then multiplied by the volts/cm indicated on the Y amplifier control knob. If the knob was set to, say, 2 V/cm, the peak-to-peak voltage of

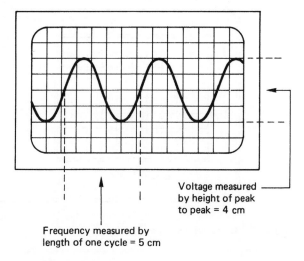

Figure 6.7 Typical trace on a CRO screen.

Fig. 6.7 would be 4 cm × 2 V/cm = 8 V. The peak voltage would be 4 V and the rms voltage 0.7071 × 4 = 2.828 V.

To measure the frequency of the waveform shown in Fig. 6.7 count the number of centimetres for one complete cycle using the one centimetre grating. The distance is shown as 5 cm in Fig. 6.7. This value is then multiplied by the time/cm on the X amplifier or time base amplifier control knob. If this knob was set to 4 ms/cm the time taken to complete one cycle would be 5 cm × 4 ms/cm = 20 ms. Frequency can be found from:

$$f = \frac{1}{T} \text{ (Hz)}$$

$$\therefore f = \frac{1}{20 \times 10^{-3}} = 1000/20 = 50 \text{ Hz}$$

The waveform shown in Fig. 6.7 therefore has an rms voltage of 2.828 V at a frequency of 50 Hz. The voltage and frequency of any waveform can be found in this way. The relevant a.c. theory is covered in Chapter 8.

EXAMPLE 1

A sinusoidal waveform is displayed on the screen of a CRO as shown in Fig. 6.7. The controls on the Y axis are set to 10 V/cm and the measurement from peak to peak is measured as 4 cm. Calculate the rms value of the waveform.

The peak-to-peak voltage is 4 cm × 10 V/cm = 40 V.
The peak voltage is 20 V. The rms voltage is

$$20 \text{ V} \times 0.7071 = 14.14 \text{ V}$$

EXAMPLE 2

A sinusoidal waveform is displayed on the screen of a CRO as shown in Fig. 6.7. The controls on the X axis are set to 2 ms/cm and the measurement for one period is calculated to be 5 cm. Calculate the frequency (f) of the waveform.

The time taken to complete one cycle (T) is 5 cm × 2 ms/cm = 10 ms.

$$f = \frac{1}{T} \text{ (Hz)}$$

$$\therefore f = \frac{1}{10 \times 10^{-3}} = 1000/10 = 100 \text{ Hz}$$

As you can see, the CRO can be used to calculate the values of voltage and frequency. It is not a *direct reading* instrument as were the analogue and digital instruments considered previously. It does, however, allow us to observe the quantity being measured unlike any other instrument and, therefore, makes a most important contribution to our understanding of electronic circuits.

Signal generators

A signal generator is an oscillator which produces an a.c. voltage of continuously variable frequency. It is used for serious electronic testing, fault-finding and experimental work. One application for a signal generator is to test the frequency response of an audio amplifier to a range of frequency.

The human ear has a frequency range of approximately 15 Hz to 15 kHz and, therefore, an audio amplifier must respond to at least this range of frequencies. This test is described in detail in Chapter 5 under Testing audio amplifiers. A signal generator is shown in Fig. 6.8.

Power supply unit (PSU)

A bench power supply unit is a very convenient way of obtaining a variable d.c. voltage from the a.c. mains. The output is very pure, a straight line when observed

Figure 6.8 A signal generator.

on a CRO, and continuously variable from zero to usually 30 V. It provides a convenient power source for bench testing or building electronic circuits. A bench power supply unit is shown in Fig. 6.9.

Figure 6.9 A bench power supply unit (PSU).

Mains electricity supply

The mains electricity supply can be lethal as all electricians and service engineers will know. It is, therefore, a sensible precaution to connect any electronic equipment being tested or repaired to a socket protected by a residual current device (RCD). Electronic equipment is protected by in-line fuses and circuit breakers and when testing suspected faulty electronic equipment, a good starting-point is to establish the presence of the mains supply. A multi-range meter with the 250 V range selected would be a suitable instrument for this purpose or, alternatively, a voltage indicator as shown in Fig. 6.10 could be used.

When isolating electronic equipment from the mains supply, in order to carry out tests or repairs, the following procedure should be followed.

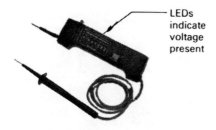

Figure 6.10 A voltage indicator.

1. Connect the voltage indicator or voltmeter to the incoming supply of the piece of equipment to be isolated. This should indicate the mains voltage and proves the effectiveness of the test instrument.
2. Isolate the supply.
3. Again test the supply to the equipment. Zero volts should be indicated on the instrument.
4. Connect the same test device to a known supply or 'proving unit' as shown in Fig. 1.10 to prove the tester is still working.
5. If the tester is proved to be working, the equipment is dead and safe to work on.

A flowchart for a secure isolation procedure is given in Fig. 1.12.

Insulation tester

The use of an insulation resistance test as described by the IEE Regulations must be avoided with any electronic equipment. The working voltage of this instrument can cause total devastation to modern electronic equipment. When carrying out an insulation resistance test as part of the prescribed series of tests for an electrical installation, all electronic equipment must first be disconnected or damage will result.

Any resistance measurements made on electronic circuits must be achieved with a battery-operated ohm meter as described previously to avoid damaging the electronic components.

Portable appliance testing (PAT)

A quarter of all serious electrical accidents involve portable electrical appliances, that is, equipment which has a cable lead and plug and which is normally moved around or can easily be moved from place to place. This includes, for example, floor cleaners, kettles, heaters, fans, televisions, desk lamps, photocopiers, fax machines and desktop computers. There is a requirement for employers under the Health and Safety at Work Act to take adequate steps to protect users of portable appliances from the hazards of electric shock and fire. The responsibility for safety applies equally to small as well as large companies. The Electricity at Work Regulations 1989 also place a duty of care upon employers to ensure that the risks associated with the use of electrical equipment are controlled.

Against this background the Health and Safety Executive (HSE) have produced guidance notes HS(G) 107 *Maintaining Portable and Transportable Electrical Equipment* and leaflets *Maintaining Portable Electrical Equipment in Offices* and *Maintaining Portable Electrical Equipment in Hotels and Tourist Accommodation*. To obtain these notes and leaflets see Appendix O for address and telephone numbers. In these publications the HSE recommend that a three level system of inspection can give cost effective maintenance of portable appliances. These are:

- user checking;
- visual inspection by an appointed person;
- combined inspection and testing by a competent person or contractor.

A **user** visually checking equipment which they are using is probably the most important maintenance procedure. About 95% of faults or damage can be identified by just looking. A user should check for obvious damage using common sense. The use of potentially dangerous equipment can then be avoided. Possible dangers to look for are as follows:

- Damage to the power cable or lead which exposes the colours of the internal conductors, which are brown, blue and green with a yellow stripe.
- Damage to the plug top itself. The plug top pushes into the wall socket, usually a square pin 13A socket in the UK, to make an electrical connection. With the plug top removed from the socket the equipment is usually electrically 'dead'. If the bakelite plastic casing of the plug top is cracked, broken or burned, or the contact pins are bent, do not use it.
- Non-standard joints in the power cable, such as taped joints.

- Poor cable retention. The outer sheath of the power cable must be secure and enter the plug top at one end and the equipment at the other. The coloured internal conductors must not be visible at either end.
- Damage to the casing of the equipment such as cracks, pieces missing, loose or missing screws or signs of melted plastic, burning, scorching or discoloration.
- Equipment which has previously been used in unsuitable conditions such as a wet or dusty environment.

If any of the above dangers are present, the equipment should not be used until the person appointed by the company to make a 'visual inspection' has had an opportunity to do so.

A **visual inspection** will be carried out by an appointed person within a company, such person having been trained to carry out this task. In addition to the user checks described above, an inspection could include the removal of the plug top cover to check that:

- a fuse of the correct rating is being used and also that a proper cartridge fuse is being used and not a piece of wire, a nail or silver paper;
- the cord grip is holding the sheath of the cable and not the coloured conductors;
- the wires (conductors) are connected to the correct terminals of the plug top as shown in Fig. 6.11;
- the coloured insulation of each conductor wire goes right up to the terminal so that no bare wire is visible;
- the terminal fixing screws hold the conductor wires securely and the screws are tight;
- all the conductor wires are secured within the terminal;
- there are no internal signs of damage such as overheating, excessive 'blowing' of the cartridge fuse or the intrusion of foreign bodies such as dust, dirt or liquids.

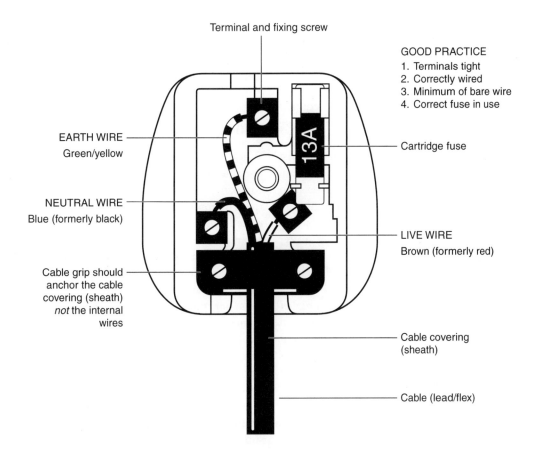

Figure 6.11 Correct connection of a plug top.

The above inspection cannot apply to 'moulded plugs', which are moulded on to the flexible cable by the manufacturer in order to prevent some of the bad practice described above. In the case of a moulded plug top, only the fuse can be checked. The visual inspection checks described above should also be applied to extension leads and their plugs. The HSE recommends that a simple procedure be written to give guidance to the 'appointed person' carrying out the visual inspection.

Combined inspection and testing is also necessary on some equipment because some faults cannot be seen by just looking – for example, the continuity and effectiveness of earth paths. For some portable appliances the earth is essential to the safe use of the equipment and, therefore, all earthed equipment and most extension leads should be periodically tested and inspected for these faults. All portable appliance test instruments (PAT Testers) will carry out two important tests, earth bonding and insulation resistance.

Earth bonding tests apply a substantial test current, typically about 25 A, down the earth pin of the plug top to an earth probe, which should be connected to any exposed metalwork on the portable appliance being tested. The PAT Tester will then calculate the resistance of the earth bond and either give an actual reading or indicate pass or fail. A satisfactory result for this test would typically be a reading of less than 0.1 Ω. The earth bond test is, of course, not required for double insulated portable appliances because there will be no earthed metalwork.

Insulation resistance tests apply a substantial test voltage, typically 500 V, between the live and neutral bonded together and the earth. The PAT Tester then calculates the insulation resistance and either gives an actual reading or indicates pass or fail. A satisfactory result for this test would typically be a reading greater than 2 MΩ.

Some PAT Testers offer other tests in addition to the two described above. These are described below.

A flash test tests the insulation resistance at a higher voltage than the 500 V test described above. The flash test uses 1.5 kV for Class 1 portable appliances, that is earthed appliances, and 3 kV for Class 2 appliances which are double insulated. The test establishes that the insulation will remain satisfactory under more stringent conditions but must be used with caution, since it may overstress the insulation and will damage electronic equipment. A satisfactory result for this test would typically be less than 3 mA.

A fuse test tests that a fuse is in place and that the portable appliance is switched on prior to carrying out other tests. A visual inspection will be required to establish that the *size* of the fuse is appropriate for that particular portable appliance.

An earth leakage test measures the leakage current to earth through the insulation. It is a useful test to ensure that the portable appliance is not deteriorating and liable to become unsafe. It also ensures that the tested appliances are not responsible for nuisance 'tripping' of RCDs (residual current devices – see Chapter 1). A satisfactory reading is typically less than 3 mA.

An operation test proves that the preceding tests were valid (i.e. that the unit was switched on for the tests), that the appliances will work when connected to the appropriate voltage supply and not draw a dangerously high current from that supply. A satisfactory result for this test would typically be less than 3.2 kW for 230 V equipment and less than 1.8 kW for 110 V equipment.

All PAT Testers are supplied with an operating manual, giving step by step instructions for their use and pass and fail scale readings. The HSE suggested intervals for the three levels of checking and inspection of portable appliances in offices and other low risk environments is given in Table 6.1.

WHO DOES WHAT?

When actual checking, inspecting and testing of portable appliances takes place will depend upon the company's safety policy and risk assessments. In low risk environments such as offices and schools, the three level system of checking, inspection and testing recommended by the HSE should be carried out. Everyone can use common sense and carry out the user checks described earlier. Visual inspections must be carried out by a 'competent person' but that person does not need to be an electrician or electronics service engineer. Any sensible member of staff who has received training can carry out this duty. They will need to know what to look for and what to do, but more importantly, they will need to be able to avoid danger to themselves and to others. The HSE recommend that the appointed person follows a simple written procedure for each visual inspection. A simple tick sheet would meet this requirement. For example:

TESTING ELECTRONIC CIRCUITS

Table 6.1 HSE suggested intervals for checking, inspecting and testing of portable appliances in offices and other low risk environments

Equipment/environment	User checks	Formal visual inspection	Combined visual inspection and electrical testing
Battery-operated: (less than 20 V)	No	No	No
Extra low voltage: (less than 50 V a.c.) e.g. telephone equipment, low voltage desk lights	No	No	No
Information technology: e.g. desktop computers, VDU screens	No	Yes, 2–4 years	No if double insulated – otherwise up to 5 years
Photocopiers, fax machines: *not* hand-held, rarely moved	No	Yes, 2–4 years	No if double insulated – otherwise up to 5 years
Double insulated equipment: *not* hand-held, moved occasionally, e.g. fans, table lamps, slide projectors	No	Yes, 2–4 years	No
Double insulated equipment: *hand-held*, e.g. some floor cleaners	Yes	Yes, 6 months–1 year	No
Earthed equipment (Class 1): e.g. electric kettles, some floor cleaners	Yes	Yes, 6 months–1 year	Yes, 1–2 years
Cables (leads) and plugs connected to the above. Extension leads (mains voltage)	Yes	Yes, 6 months–4 years depending on the type of equipment it is connected to	Yes, 1–5 years depending on the type of equipment it is connected to

1 Is the correct fuse fitted? Yes/No
2 Is the cord grip holding the cable sheath? Yes/No

The tick sheet should incorporate all the appropriate visual checks and inspections described earlier.

Testing and inspection require a much greater knowledge than is required for simple checks and visual inspections. This more complex task need not necessarily be carried out by a qualified electrician or electronics service engineer. However, the person carrying out the test must be trained to use the equipment and to interpret the results. Also, greater knowledge will be required for the inspection of the range of portable appliances which might be tested.

KEEPING RECORDS

Records of the inspecting and testing of portable appliances are not required by law but within the Electricity at Work Regulations 1989, it is generally accepted that some form of recording of results is required to implement a quality control system. The control system should:

- ensure that someone is nominated to have responsibility for portable appliance inspection and testing;
- maintain a log or register of all portable appliance test results to ensure that equipment is inspected and tested when it is due;
- label tested equipment with the due date for its next inspection and test as shown in Fig. 6.12.

Figure 6.12 Typical PAT Test labels.

Any piece of equipment which fails a PAT Test should be disabled and taken out of service (usually by cutting off the plug top), labelled as faulty and sent for repair.

The register of PAT Test results will help managers to review their maintenance procedures and the frequency of future visual inspections and testing. Combined inspection and testing should be carried out where there is a reason to suspect that the equipment may be faulty, damaged or contaminated but cannot be verified by visual inspection alone. Inspection and testing should also be carried out after any repair or modification to establish the integrity of the equipment or at the start of a maintenance system, to establish the initial condition of the portable equipment being used by the company.

7

DIGITAL ELECTRONICS

Digital electronics embraces all of today's computer-based systems. These are decision-making circuits which use what is known as combinational logic in applications such as industrial robots, industrial hydraulic and pneumatic systems, programmable logic controllers, telephone exchanges, motor vehicle and domestic appliance control systems, children's toys and their parents' personal computers and audio equipment. Digital electronics is concerned with straightforward two-state switching circuits. The simplicity and reliability of this semiconductor transistor switching has encouraged designers to look for new digital markets. Traditional applications which have analogue inputs, such as audio recordings, are now using digital techniques, with the development of analogue-to-digital converters. These convert the analogue voltage signals into digital numbers. A digital and analogue waveform are shown in Fig. 7.1.

The digital waveform has two quite definite states, either on or off, and changes between these two states very rapidly. An analogue waveform changes its value smoothly and progressively between two extremes.

In an analogue system, changes in component values due to ageing and temperature can affect the circuit's performance. Digital systems are very much less susceptible to individual component changes. Another significant advantage of digital circuits is their immunity to noise and interference signals. With analogue circuitry this is a nuisance, particularly when signal levels are very small and, therefore, easily contaminated by noise. Digital signals, however, have a very large amplitude and can, therefore, be made relatively free of noise which helps manufacturers to achieve a very high quality of sound reproduction, as anyone who listens to a compact disc recording can testify. Logic circuits have been developed to deal with these digital, two-state switching circuits. Information is expressed as *binary numbers*, that is numbers which consist of ones and zeros. These two binary states are represented by low and high voltages, where low voltage is 0 V and high voltage is say +5V. The low level is called Logic 0 and the high level Logic 1. When the voltage level of a digital signal is not rapidly changing it remains steady at one of these two levels. Information is processed according to rules built into circuits made up of single units called *logic gates*. Each unit is called a *gate*, because like a gate it can allow information to pass through or stop it, and *logic* gate because it behaves according to rules which can be described by logical or predictable statements.

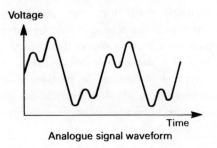
Analogue signal waveform

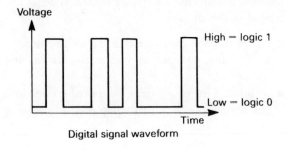
Digital signal waveform

Figure 7.1 Comparison of an analogue and digital waveform.

A logic gate may have a number of inputs but has only one output which can only be either logic 1 or logic 0, no other value exists. The basic range of logic gates is known by the names AND, OR, NOT, NOR and NAND.

The AND logic gate

The operation of this gate can probably best be understood by drawing a simple switch equivalent circuit, as shown in Fig. 7.2. The logic symbol is also shown. The signal lamp will only illuminate if switch A *and* switch B are closed, or we could say the output F of the gate will only be at logic 1 if input A *and* input B are both at logic 1.

If the AND gate was operating a car handbrake warning lamp, it would only illuminate when the handbrake *and* the ignition was on. The *truth table* shows the output state for all possible combinations of inputs.

The OR gate

The OR gate can be represented by parallel connected switches, as shown in Fig. 7.3 which also shows the logic symbol. In this case the signal lamp will only illuminate if switch A *or* switch B *or* both switches are closed. Alternatively, we could say that the output F will only be at logic 1 if input A *or* input B *or* both inputs are at logic 1.

If the OR gate was operating an interior light in a motor car, it would illuminate when the nearside door was opened *or* the offside door was opened *or* when both doors were opened. The truth table shows the output state for all possible combinations of inputs.

The Exclusive-OR gate

The Exclusive-OR gate is an OR gate with only two inputs which will give a logic 1 output only if input A *or* input B is at logic 1, but *not* when both A and B are

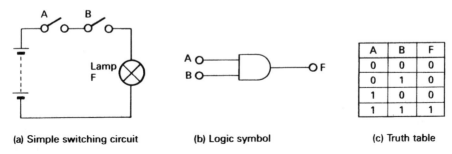

(a) Simple switching circuit   (b) Logic symbol   (c) Truth table

Figure 7.2 The AND gate.

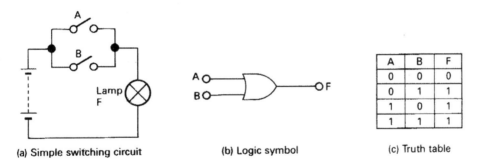

(a) Simple switching circuit   (b) Logic symbol   (c) Truth table

Figure 7.3 The OR gate.

(a) Logic symbol

A	B	F
0	0	0
0	1	1
1	0	1
1	1	0

(b) Truth table

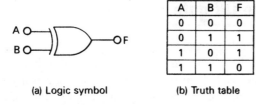

Figure 7.4 The Exclusive -OR gate.

at logic 1. The symbol and truth table are given in Fig. 7.4

The NOT gate

The NOT gate is a single input gate which gives an output that is the opposite of the input. For this reason it is sometimes called an *inverter* or a *negator* or simply a *sign changer.* If the input is A, the output is *not* A which is written as $\bar{A}$ (bar A). The small circle on the output of the gate always indicates a change of sign.

If the NOT gate was operating a spin dryer motor it would only allow the motor to run when the lid was *not* open. The truth table shows the output state for all possible inputs in Fig. 7.5.

(a) Logic symbol

INPUT	OUTPUT
0	1
1	0

(b) Truth table

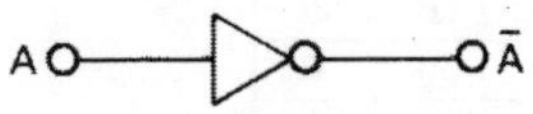

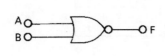

Figure 7.5 The NOT gate.

The NOR gate

The NOR gate is a NOT gate and an OR gate combined to form a NOT-OR gate. The output of the NOR gate is the opposite of the OR gate as can be seen by comparing the truth table for the NOR gate in Fig. 7.6 with that of the OR gate.

A	B	F
0	0	1
0	1	0
1	0	0
1	1	0

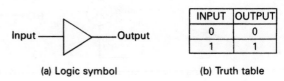

Figure 7.6 The NOR gate.

The NAND gate

The NAND gate is a NOT gate and an AND gate combined to form a NOT-AND gate. The output of the NAND gate is the opposite of the AND gate as can be seen by comparing the truth table for the NAND gate in Fig. 7.7 with that of the AND gate.

(a) Logic symbol

A	B	F
0	0	1
0	1	1
1	0	1
1	1	0

(b) Truth table

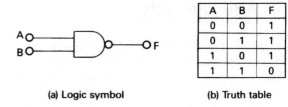

Figure 7.7 The NAND gate.

Buffers

The simplest of all logic devices is the buffer. This device has only one input and one output, and its logical output is exactly the same as its logical input. Given that this device has no effect upon the logic levels within a circuit, you may be wondering what the purpose of such an apparently redundant device might be! Well, although the input and output voltage levels of the buffer are identical, the *currents* present at the input and output can be *very* different. The output current can be much greater than the input current and, therefore, buffers can be said to exhibit *current gain.* In this way, buffers can be used to interface logic circuits to other circuits which demand more current than could be supplied by an unbuffered logic circuit. The symbol used to represent a buffer is shown in Fig. 7.8.

(a) Logic symbol

INPUT	OUTPUT
0	0
1	1

(b) Truth table

Figure 7.8 The buffer.

Logic networks

Individual logic gates may be interconnected to provide any desired output. The results of any combination can be found by working through each individual gate in the combination or logic system in turn, and producing the truth table for the particular network. It can also be very instructive to build up logic gate combinations on a logic simulator and to confirm the theoretical results. This facility will undoubtedly be available if the course of study is being undertaken at a Technical college, Training Centre or Evening Institute.

EXAMPLE 1

Two logic gates are connected together as shown in Fig. 7.9. Complete the truth table for this particular logic network. In considering Fig. 7.9 and working as always from left to right, we can see that an AND gate feeds a NOT gate. The whole network has two inputs, A and B, and one output F. The first step in constructing the truth table for the combined logic gates is to label the outputs of *all* the gates and prepare a blank truth table as shown in Fig. 7.9. Let us call the output of the AND gate C (it could be any letter except A, B or F) and work our way progressively through the individual gates from left to right. For any two input logic gates, there are four possible combinations, 00, 01, 10 and 11. When these are included on the truth table it will appear as shown in Fig. 7.10. The next step is to complete column C. Now, C is the output of an AND gate and can, therefore, only be at logic 1 when both A *and* B are logic 1. The truth table can, therefore, be completed as shown in Fig. 7.11. The final step is to complete column F, the output of a NOT gate whose input combinations are given by column C. A NOT gate is a single input gate whose output is the opposite of the input and, therefore, the output column F must be the opposite of column C, as shown by Fig. 7.12. The truth table tells us that this particular combination of gates will give a logic 1 output with any input combination *except* when A and B are both at logic 1. This combination, therefore, behaves like a NAND gate as can be confirmed by referring to Fig. 7.7.

A	B	C	F
0	0	0	
0	1	0	
1	0	0	
1	1	1	

Figure 7.11 Truth table for Example 1.

A	B	C	D
0	0	0	1
0	1	0	1
1	0	0	1
1	1	1	0

Figure 7.12 The completed truth table for Example 1.

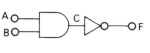

(a) Logic network   (b) Blank truth table

Figure 7.9 Logic network and blank truth table for Example 1.

A	B	C	F
0	0		
0	1		
1	0		
1	1		

Figure 7.10 Truth table for Example 1.

EXAMPLE 2

A NAND and NOT gate are connected together as shown in Fig. 7.13. Complete a truth table for this particular network. The truth table for this particular combination can be constructed in exactly the same way as for Example 1. The NAND gate has two inputs P and Q and an output R. The NOT gate has an input R and output S.

All possible combinations of inputs are shown in columns P and Q of the truth table shown in Fig. 7.14. A NAND gate will give a logic 1 output for *all* combinations of inputs *except* when input A *and* B are at logic

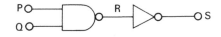

Figure 7.13 Logic network for Example 2.

P	Q	R	S
0	0	1	0
0	1	1	0
1	0	1	0
1	1	0	1

Figure 7.14 Truth table for Example 2.

1 as shown by column R of Fig. 7.14. The second and final gate in this network is a NOT gate which provides an output which is the reverse of the input. The output, given by column S of the truth table, will therefore be the reverse of column R as shown in Fig. 7.14.

This particular combination will, therefore, give a logic 1 output only when input P *and* input Q are at logic 1. Therefore, it can be seen that the combination of a NAND and a NOT gate produces the equivalent of an AND gate. This can be checked by referring back to Fig. 7.2.

EXAMPLE 3

A NAND gate has a NOT gate on each of its inputs as shown in Fig. 7.15. Construct a truth table for this particular network. The NOT gates will invert or reverse the input. We can, therefore, call the output of these NOT gates, not A and not B, which is written as $\bar{A}$ and $\bar{B}$. This then provides the input to the NAND gate. A NAND gate will provide a logic 1 output for any input combination *except* when both inputs are at logic 1. The truth table can, therefore, be developed as shown in Fig. 7.16. It can

be seen by referring back to Fig. 7.3, and comparing the inputs A and B and output F, that this combination gives the network equivalent of an OR gate. That is, the output is at logic 1 if the input A *or* input B *or* both are at logic 1.

EXAMPLE 4

A logic network is assembled as shown in Fig. 7.17. Develop a truth table and describe in a sentence the relationship between the input and output. The truth table for this particular combination can be drawn up as shown in Fig. 7.18. There are only two inputs A and B. The output C of the AND gate and the output D of the OR gate provide the input to a NOR gate, which provides the output F.

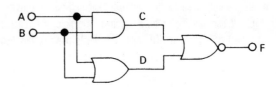

Figure 7.17 Logic network for Example 4.

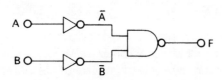

Figure 7.15 Logic network for Example 3.

A	B	$\bar{A}$	$\bar{B}$	F
0	0	1	1	0
0	1	1	0	1
1	0	0	1	1
1	1	0	0	1

Figure 7.16 Truth table for Example 3.

A	B	C	D	F
0	0	0	0	1
0	1	0	1	0
1	0	0	1	0
1	1	1	1	0

Figure 7.18 Truth table for Example 4.

The output of an AND gate is high, that is at logic 1, only when input A *and* input B are at logic 1. Column C of the truth table shows the output of the AND gate for all combinations of input. The output of an OR gate is high, when input A *or* input B *or* both are high. This is shown by column D of the truth table. The input to the final NOR gate is provided by the logic levels indicated in columns C and D and the output F is, therefore, as shown in column F. The output of this combination of logic gates is high, that is at logic 1, only when input A *and* input B are low. This is equivalent to a single NOR gate.

In the examples considered until now, the inputs have been restricted to only two variables. In practice, logic gates may be constructed with many inputs and the truth tables developed as shown above. However, when there are more than three inputs the truth table becomes very cumbersome because the number of lines required for the truth table follows the law of 2^n where n is equal to the number of inputs. Therefore, a two-input gate requires 2^2 (4) lines, as can be seen in the previous examples, a three-input gate 2^3 (8) lines, a four-input gate 2^4 (16) lines etc.

EXAMPLE 5

A logic system having three inputs is assembled as shown in Fig. 7.19. Develop a truth table and describe in a sentence the relationship between the input and output. The truth table for this combination of logic gates can be drawn up as shown in Fig. 7.20. Three inputs mean that the truth table must have 2^3 rows, that is eight rows. All possible combinations of input are shown in columns A, B and C. The first AND gate will give a logic 1 output only when input A and B are both logic 1. There are two such occasions as shown by column D. The second AND gate will give a logic 1 output only when input C and D are both logic 1. This occurs on only one occasion. That is, the output is at logic 1 only when all three inputs are at logic 1.

EXAMPLE 6

A three input logic network is assembled as shown in Fig. 7.21. Develop a suitable truth table and use this to describe the relationship between the three inputs and the output Z. The truth table for this network, which has three inputs can be constructed as shown in Fig. 7.22. All possible combinations of the input are shown in columns V, W and X.

The first OR gate will give a logic 1 output when either V or W or both are at logic 1. This occurs on all but two occasions as can be seen by considering column Y of the truth table. The second OR gate will give a logic 1 output when either X or Y or both are at logic 1. This occurs on all but one occasion. Therefore, we can say that the output Z is at logic 0 only when all three inputs are at logic 0. If any input is at logic 1, the input Z is also at logic 1.

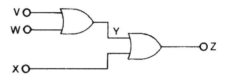

Figure 7.21 Logic network for Example 6.

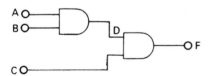

Figure 7.19 Logic network for Example 5.

V	W	X	Y	Z
0	0	0	0	0
0	0	1	0	1
0	1	0	1	1
0	1	1	1	1
1	0	0	1	1
1	0	1	1	1
1	1	0	1	1
1	1	1	1	1

Figure 7.22 Truth table for Example 6.

A	B	C	D	F
0	0	0	0	0
0	0	1	0	0
0	1	0	0	0
0	1	1	0	0
1	0	0	0	0
1	0	1	0	0
1	1	0	1	0
1	1	1	1	1

Figure 7.20 Truth table for Example 5.

Logic families

The simplicity of digital electronics with its straightforward on off switching means that many logic elements can be packed together in a single integrated circuit and packaged as a standard dual-in-line IC as shown in Figs. 4.19 and 21. Different types of semiconductor circuitry can be used to construct the logic gates. Each type is called a logic family because all members of that integrated circuit family will happily work together in a circuit.

Two main families of digital logic have emerged as the most popular with designers of general-purpose digital circuits in recent years. These are the TTL and CMOS families. The older of these is the TTL (transistor-transistor logic) family which was introduced in 1964 by Texas Instruments Ltd. The standard TTL family is designated the 7400 series. Figure 7.23 shows the internal circuitry of a TTL 7400 IC. This contains a quad 2-input NAND gate, that is, it contains four NAND gates each with two inputs and one output. Thus, with two power supply connections, the 7400 IC has 14 connections and is manufactured as the familiar 14 pin dual-in-line package. Many other combinations are available and each has its own unique number which, in this family, always begins with 74 and is followed by two other numbers. The final two numbers indicate the type of logic gate, for example, a 7432 is a quad 2-input OR gate, a 7411 a triple 3-input AND gate, as can be seen from the data sheets given in Appendix L.

The CMOS family, pronounced see-mos, is the Complementary Metal Oxide Semiconductor family of logic ICs which was introduced in 1968. The best known CMOS family is designated the 4000 series and, like its TTL equivalent, is housed in a 14 pin dual-in-line package. The 4011B is a quad 2-input NAND gate, as shown in Fig. 7.24. This is *similar* to the TTL 7400 shown in Fig. 7.23 but it is not *identical* because the pin connections differ and, therefore, a TTL package cannot replace a CMOS package.

The theory of digital logic is the same for all logic families. The differences between the families are confined to the practical aspects of the circuit design. Each logic family has its own special characteristics which make it appropriate for particular applications.

Comparison of TTL and CMOS

A CMOS device dissipates about 1 mW per logic gate compared with about 20 mW for a standard TTL logic gate. Therefore, CMOS has a much lower power consumption than TTL which is particularly important when the circuitry is to be battery powered.

The output of a logic gate may be connected to the input of many other logic gates. The drive capability of a gate to hold its input at logic 0 or logic 1 while delivering current to the other gates in the circuit is called the *fan-out* capability. The fan-out for for TTL is ten, which means that ten other TTL logic gates can take their input from one TTL output and still switch reliably before overloading occurs. A fan-out of fifty is typical for CMOS because they have a very high input impedance and low power consumption.

The power supply for TTL must be 5 V ± 0.25 V with a ripple of less than 5% peak to peak. A TTL device will be damaged if voltages in excess of these limits are applied. This requirement can be easily

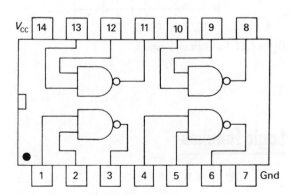

Figure 7.23 The 7400 TTL logic family.

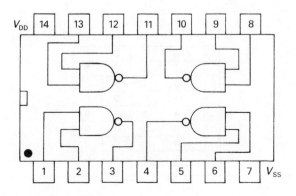

Figure 7.24 The 4011B CMOS logic family.

met by the IC fixed voltage regulators discussed in Chapter 7. CMOS devices can tolerate a much wider variation of supply voltages, typically +3 V to +15 V.

Another advantage of CMOS logic circuits is that they require only about one fiftieth of the 'floor space' on a silicon chip compared with TTL. CMOS is, therefore, ideal for complex silicon chips such as those required by microprocessors and memories.

The switching times of any logic network are infinitesimal when compared with an electromechanical relay. However, the switching times for TTL logic are very much faster than CMOS, although both are measured in nanoseconds (10^{-9} s). The properties of each family are summarised in Table 7.1.

Table 7.1 Properties of logic families

Property	TTL	CMOS
Power consumption	high—20 mW	low—1 mW
Operating current	high—mA range	low—µA range
Power supply	5 V ± 0.25 V d.c.	3V to 15 V d.c.
Switching speeds	fast—10 ns	slow—100 ns
Input impedance	low	high
Fan-out	10	50

Working with logic

ICs of the same number will always have the same function regardless of the manufacturer and any suffix or prefix which may accompany the basic gate number. Therefore, an IC package must be replaced with another of the same number. The very high input impedance of CMOS accounts for its low power consumption but it does mean that static electricity can build up on the input pins if they come into contact with plastic, nylon or the manmade fibres of workers' clothing during circuit assembly or repair. This does not happen with TTL because the low input impedance ensures that any static charges leak harmlessly away through the junctions in the IC. Static voltages on CMOS can destroy them, and they are supplied with anti-static carriers and these should not be removed until wiring is completed. Internal protection is also provided by buffered inputs but these cannot become effective until the supply is connected. Inputs must, therefore, be disconnected before the mains connections when disconnecting CMOS. Alternatively, the power supplies must be connected before the inputs when assembling CMOS chips. Input signals must not be applied until the power supply is connected and switched on.

When operating CMOS with normal positive logic signals VSS is the common line (OV) and VDD is the positive connection, 3 to 15 volts. Unused inputs must not be left *floating*. They must always be connected in parallel with similar used inputs, or connected to the supply rail.

Working with CMOS has created many new problems for electronic technicians. These can be overcome by:

- working on a copper plate working surface which is connected to earth;
- ensuring that all equipment is properly earthed;
- wearing a conductive wrist-band which is connected to the earth of the working surface.

When these precautions are observed the problems of handling CMOS ICs can be overcome without too much difficulty.

British Standard symbols

Although the British Standards recommend symbols for logic gates, much of the manufacturers' information uses the American 'MilSpec' Standard symbols. For this reason I have reluctantly used the American standard symbols in this chapter. However, there is some pressure in the UK to adopt the BS symbols and for this reason the British Standard and American Standard symbols are cross-referenced in Appendix K.

Exercises

1 A voltage signal which changes smoothly and progressively between two extremes is called
   (a) a logical waveform
   (b) an analogue waveform
   (c) an interference signal
   (d) a digital waveform

2 A voltage signal which has two quite definite states, either on or off, is called
   (a) a logical waveform
   (b) an analogue waveform
   (c) an interference signal
   (d) a digital waveform
3 A single logic gate has two inputs X and Y and one output Z. The output Z will be at logic 1 only when input A and input B are at logic 1 if the gate is
   (a) a NOT gate
   (b) an AND gate
   (c) an OR gate
   (d) a NOR gate
4 Develop the truth table for an Exclusive-OR gate.
5 Develop the truth table for a NOR gate.
6 For the circuit shown in Fig. 7.25 develop the truth table.

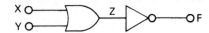

Figure 7.25 Logic network for question number 6.

7 Develop the truth table for the network shown in Fig. 7.26 and describe the relationship between the inputs and output.

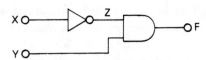

Figure 7.26 Logic network for question number 7.

8 Develop the truth table for the logic system shown in Fig. 7.27 and describe the relationship between the output and inputs.

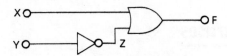

Figure 7.27 Logic network for question number 8.

9 Work out the truth table for the circuit shown in Fig. 7.28. Describe in a sentence the behaviour of this circuit.

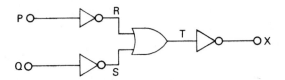

Figure 7.28 Logic network for question number 9.

10 Complete the truth table for the circuit shown in Fig. 7.29 and describe the circuit behaviour.

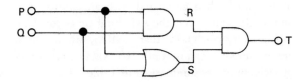

Figure 7.29 Logic network for question number 10.

11 Using a truth table describe the output of the logic system shown in Fig. 7.30.

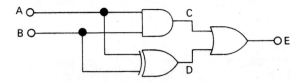

Figure 7.30 Logic network for question number 11.

12 Use a truth table to describe the output of the logic network shown in Fig. 7.31.

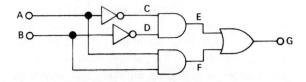

Figure 7.31 Logic network for question number 12.

13 Inputs A, B and C of Fig. 7.32 are controlled by three separate key switches. Determine the sequence of key switch positions which will give an output at F.

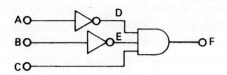

Figure 7.32 Logic network for question number 13.

ELECTRONIC CIRCUIT THEORY

This chapter brings together most of the circuit theory in the approved training courses which electricians and service engineers undertake. Electricians and service engineers should, therefore, consider this chapter as revision while readers from other disciplines might like to 'dip into' the theory on a 'need to know' basis.

Units

Very early units of measurement were based on the things easily available – length of a stride, the distance from the nose to the outstretched hand, the weight of a stone and the time-lapse of one day. Over the years, new units were introduced and old ones were modified. Different branches of science and engineering were working in isolation, using their own units, and the result was an overwhelming variety of units.

In all branches of science and engineering there is a need for a practical system of units which everyone can use. In 1960 the General Conference of Weights and Measures agreed to an international system called the Système International d'Unités (abbreviated to SI units). SI units are based upon a small number of fundamental units from which all other units may be derived, see Table 8.1.

Like all metric systems, SI units have the advantage that prefixes representing various multiples or sub-multiples may be used to increase or decrease the size of the unit by various powers of ten. Some of the more common prefixes and their symbols are shown in Table 8.2.

Basic circuit theory

All matter is made up of atoms which arrange themselves in a regular framework within a material. The atom is made up of a central, positively charged

Table 8.1 SI units

SI Unit	Measure of	Symbol
The fundamental units		
Metre	Length	m
Kilogram	Mass	kg
Second	Time	s
Ampere	Electric current	A
Kelvin	Thermodynamic temperature	K
Candela	Luminous intensity	cd
Some derived units		
Coulomb	Charge	C
Joule	Energy	J
Newton	Force	N
Ohm	Resistance	Ω
Volt	Potential difference	V
Watt	Power	W

Table 8.2 Prefixes for use with SI units

Prefix	Symbol	Multiplication factor		
mega	M	$\times 10^6$	or	$\times 1\,000\,000$
kilo	k	$\times 10^3$	or	$\times 1000$
hecto	h	$\times 10^2$	or	$\times 100$
deca	da	$\times 10$	or	$\times 10$
deci	d	$\times 10^{-1}$	or	$\div 10$
centi	c	$\times 10^{-2}$	or	$\div 100$
milli	m	$\times 10^{-3}$	or	$\div 1000$
micro	μ	$\times 10^{-6}$	or	$\div 1\,000\,000$

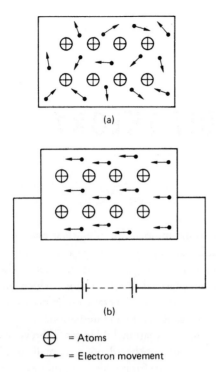

Figure 8.1 Atoms and electrons on a material.

nucleus, surrounded by negatively charged electrons. The electrical properties of a material depend largely upon how tightly these electrons are bound to the central nucleus.

A *conductor* is a material in which the electrons are loosely bound to the central nucleus and are, therefore, free to drift around the material at random from one atom to another as shown in Fig. 8.1(a). Materials which are good conductors include copper, brass, aluminum and silver.

An *insulator* is a material in which the outer electrons are tightly bound to the nucleus and so there are no free electrons to move around the material. Good insulating materials are PVC, rubber, glass and wood.

If a battery is attached to a conductor as shown in Fig. 8.1(b), the free electrons drift purposefully in one direction only. The free electrons close to the positive plate of the battery are attracted to it, since unlike charges attract, and the free electrons near the negative plate will be repelled from it. For each electron entering the positive terminal of the battery, one will be ejected from the negative terminal, so the number of electrons in the conductor remains constant.

This drift of electrons within a conductor is known as an electric *current*, measured in amperes and given the symbol I. For a current to continue to flow, there must be a complete circuit for the electrons to move around. If the circuit is broken by opening a switch, for example, the electron flow, and therefore the current, will stop immediately.

To cause a current to flow continuously around a circuit, a driving force is required, just as a circulating pump is required to drive water round a central heating system. This driving force is the *electromotive force* (abbreviated to emf) and is the energy which causes the current to flow in a circuit. Each time an electron passes through the source of emf, more energy is provided to send it on its way around the circuit.

An emf is always associated with energy conversion, such as chemical to electrical in batteries and mechanical to electrical in generators. The energy introduced into the circuit by the emf is transferred to the load terminals by the circuit conductors. The *potential difference* (abbreviated to p.d.) is the change in energy levels measured across the load terminals. This is also called the volt drop or terminal voltage, since emf and p.d. are both measured in volts. Every circuit offers some opposition to current flow which we call the circuit *resistance* measured in ohms, to commemorate the famous experimenter George Simon Ohm, who was responsible for the analysis of electrical circuits. The symbol Ω represents an ohm.

In 1826, Ohm published details of an experiment he had made to investigate the relationship between the current passing through and the potential difference between the ends of wire. As a result of this experiment he arrived at the following law, known as *Ohm's law*: 'The current passing through a conductor under constant temperature conditions is proportional to the potential difference across the conductor'.

$$V = I \times R \ (V)$$

Transposing this formula, we also have

$$I = \frac{V}{R} \ (A) \text{ and } R = \frac{V}{I} \ (\Omega)$$

ELECTRONIC CIRCUIT THEORY

EXAMPLE 1

An electric heater when connected to a 230 V supply was found to take a current of 4 A. Calculate the element resistance.

$$R = \frac{V}{I} \ (\Omega)$$

$$\therefore R = \frac{230 \text{ V}}{4 \text{ A}} = 57.5 \ \Omega$$

EXAMPLE 2

The insulation resistance measured between phase conductors on a 400 V supply was found to be 2 MΩ. Calculate the leakage current.

$$I = \frac{V}{R} \ (A)$$

$$\therefore I = \frac{400}{2 \times 10^6 \ \Omega} = 200 \times 10^{-6} \text{ A} = 200 \ \mu A$$

EXAMPLE 3

When a 4Ω resistor was connected across the terminals of an unknown d.c. supply, a current of 3A flowed. Calculate the supply voltage.

$$V = I \times R \ (V)$$

$$\therefore V = 3A \times 4\Omega = 12V$$

Resistivity (symbol ρ – the Greek letter 'rho')

The resistance or opposition to current flow varies for different materials, each having a particular constant value. If we know the resistance of say one metre of a material, then the resistance of five metres will be five times the resistance of one metre.

The *resistivity* of a material is defined as the resistance of a sample of unit length and unit cross-section. Typical values are given in Table 8.3. Using the constants for a particular material we can calculate the resistance of any length and thickness of that material from the equation

$$R = \frac{\rho l}{a} \ (\Omega)$$

where ρ = the resistivity constant for the material (Ωm)
l = the length of the material (m)
a = the cross-sectional area of the material (m²)

Table 8.3 gives the resistivity of silver as 16.4×10^{-9} ohm metre which means that a sample of silver one metre long and one metre in cross-section will have a resistance of $16.4 \times 10^{-9} \ \Omega$.

Table 8.3 Resistivity values

Material	Resistivity [ohm metre]
Silver	16.4×10^{-9}
Copper	17.5×10^{-9}
Aluminium	28.5×10^{-9}
Brass	75.0×10^{-9}
Iron	100.0×10^{-9}

EXAMPLE 1

Calculate the resistance of 100 metres of copper cable of 1.5 mm² cross-sectional area if the resistivity of copper is taken as $17.5 \times 10^{-9} \ \Omega$m.

$$R = \frac{\rho l}{a} \ (\Omega)$$

$$\therefore R = \frac{17.5 \times 10^{-9} \ \Omega \text{m} \times 100 \text{ m}}{1.5 \times 10^{-6} \text{ m}^2} = 1.16 \ \Omega$$

EXAMPLE 2

Calculate the resistance of 100 metres of aluminium cable of 1.5 mm² cross-sectional area if the resistivity of aluminium is taken as $28.5 \times 10^{-9} \ \Omega$m.

$$R = \frac{\rho l}{a} \ (\Omega)$$

$$\therefore R = \frac{28.5 \times 10^{-9} \ \Omega \text{m} \times 100 \text{ m}}{1.5 \times 10^{-6} \text{ m}^2} = 1.9 \Omega$$

The above examples show that the resistance of an aluminium cable is some 60% greater than a copper conductor of the same length and cross-section. Therefore, if an aluminium cable is to replace a copper cable, the conductor size must be increased to carry the rated current as given by the tables of the IEE Regulations (BS7671).

The three effects of an electric current

When an electric current flows in a circuit it can have one or more of the following three effects: *heating*, *magnetic* or *chemical*.

HEATING EFFECT

The movement of electrons within a conductor, which is the flow of an electric current, causes an increase in the temperature of the conductor. The amount of heat generated by this current flow depends upon the type and dimensions of the conductor and the quantity of current flowing. By changing these variables, a conductor may be operated hot and used as the heating element of a fire, or be operated cool and used as an electrical installation conductor.

The heating effect of an electric current is also the principle upon which a fuse gives protection to a circuit. The fuse element is made of a metal with a low melting point and forms a part of the electrical circuit. If an excessive current flows, the fuse element overheats and melts, breaking the circuit.

MAGNETIC EFFECT

Whenever a current flows in a conductor a magnetic field is set up around the conductor like an extension of the insulation. This is further discussed later in this chapter under the subheading 'magnetism and motors'. The magnetic field increases with the current and collapses if the current is switched off. A conductor carrying current and wound into a solenoid produces a magnetic field very similar to a permanent magnet, but has the advantage of being switched on and off by any switch which controls the circuit current.

The magnetic effect of an electric current is the principle upon which electric bells, relays, instruments, motors and generators work.

CHEMICAL EFFECT

When an electric current flows through a conducting liquid the liquid is separated into its chemical parts. The conductors which make contact with the liquid are called the anode and cathode. The liquid itself is called the electrolyte, and the process is called *electrolysis*.

Electrolysis is an industrial process used in the refining of metals and electroplating. It was one of the earliest industrial applications of an electric current. Most of the aluminum produced today is extracted from its ore by electrochemical methods. Electroplating serves a double purpose by protecting a base metal from atmospheric erosion and also giving it a more expensive and attractive appearance. Silver and nickel plating has long been used to enhance the appearance of cutlery, candlesticks and sporting trophies.

An anode and cathode of dissimilar metals placed in an electrolyte can react chemically and produce an emf. When a load is connected across the anode and cathode, a current is drawn from this arrangement which is called a cell. A battery is made up of a number of cells. It has many useful applications in providing portable electrical power, but electrochemical action can also be undesirable since it is the basis of electrochemical corrosion which rots our motor cars, industrial containers and bridges.

Electrostatics

If a battery is connected between two insulated plates, the emf of the battery forces electrons from one plate to another until the p.d. between the plates is equal to the battery emf.

The electrons flowing through the battery constitute a current, I, amperes which flows for a time, t, seconds. The plates are then said to be charged. The amount of charge transferred

$$Q = It \, (C)$$

Figure 8.2 shows the charges on a capacitor's plates.

When the voltage is removed the charge Q is trapped on the plates, but if the plates are joined together, the same quantity of electricity, $Q = It$, will flow back from one plate to the other, so discharging them. The property of two plates to store an electric charge is called its *capacitance*.

By definition, a capacitor has a capacitance of one farad when a p.d. of one volt maintains a charge of one coulomb on that capacitor, or

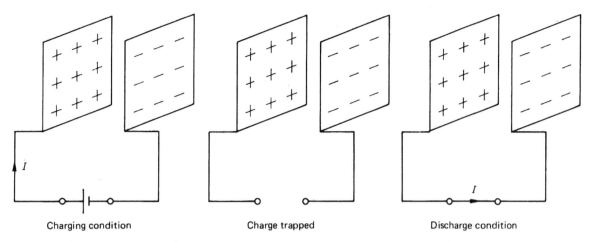

Figure 8.2 The charge on a capacitor's plates.

$$C = \frac{Q}{V} \text{ (F)}$$

Collecting these important formulae together we have

$$Q = It = CV$$

Capacitors

A capacitor consists of two metal plates, separated by an insulating layer called the dielectric. It has the ability of storing a quantity of electricity as an excess of electrons on one plate and a deficiency on the other.

EXAMPLE 1

A 100 µF capacitor is charged by a steady current of 2 mA flowing for five seconds. Calculate the total charge stored by the capacitor and the p.d. between the plates.

$$Q = It \text{ (C)}$$
$$= 2 \times 10^{-3} \text{ A} \times 5\text{s} = 10 \text{ mC}$$
$$Q = CV$$
$$\therefore V = \frac{Q}{C} \text{ (V)}$$
$$= \frac{10 \times 10^{-3} \text{ C}}{100 \times 10^{-6} \text{ F}} = 100 \text{ V}$$

The p.d. which may be maintained across the plates of a capacitor is determined by the type and thickness of the dielectric medium. Capacitor manufacturers usually indicate the maximum safe working voltage for their products.

Capacitors are classified by the type of dielectric material used in their construction. Chapter 2 shows the general construction and appearance of capacitors to be found in electronic work.

CHARGING CAPACITORS

Connecting a voltage to the plates of a capacitor causes it to charge up to the potential of the supply. This involves electrons moving around the circuit to create the necessary charge conditions and, therefore, this action does not occur instantly, but takes some time. This scientific fact has many applications in electronic circuits.

C-R CIRCUITS

Figure 8.4 shows the circuit diagram for a simple C-R circuit and the graphs drawn from the meter readings. It can be seen that:

- initially the current has a maximum value and decreases slowly to zero as the capacitor charges;
- initially the capacitor voltage rises rapidly but then slows down, increasing gradually until the capacitor voltage is equal to the supply voltage when fully charged.

The mathematical name for the shape of these curves is an *exponential* curve and, therefore, we say that the

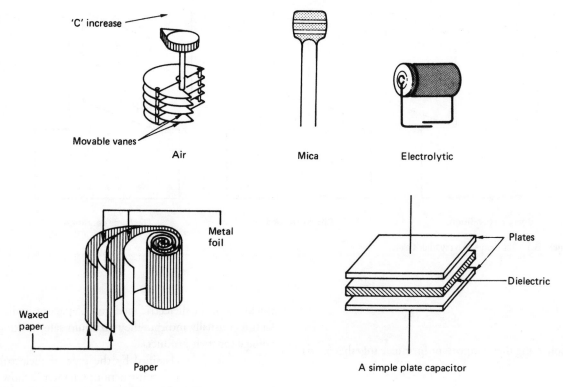

Figure 8.3 Construction and appearance of capacitors.

capacitor voltage is growing exponentially while the current is decaying exponentially during the charging period. The *rate* at which the capacitor charges is dependent upon the *size* of the capacitor and resistor. The bigger the values of C and R, the longer will it take to charge the capacitor. The time taken to charge a capacitor by a *constant* current is given by the *time constant* of the circuit which is expressed mathematically as

$$T = CR \text{ (s)}$$

where T is the time in seconds.

EXAMPLE 1

A 60 µF capacitor is connected in series with a 20 kΩ resistor across a 12V supply. Determine the time constant of this circuit.

$$T = CR \text{ (s)}$$
$$\therefore T = 60 \times 10^{-6} \text{ F} \times 20 \times 10^{3} \text{ Ω}$$
$$= 1.2 \text{ (s)}$$

We have already seen that in practice the capacitor is not charged by a *constant* current but, in fact, charges exponentially. However, it can be shown by experiment that in *one* time constant the capacitor will have

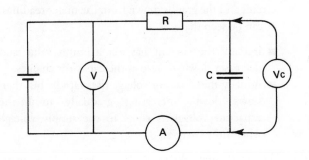

Figure 8.4 A C-R circuit.

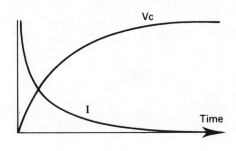

reached about 63% of its final steady value, taking about five times the time constant to become fully charged. Therefore, in 1.2 s the 60μF capacitor of Example 1 will have reached about 63% of 12 V and after 5 T, that is 6 s, will be fully charged at 12 V.

Resistors

In an electrical or electronic circuit, resistors may be connected in series, in parallel or in various combinations of series and parallel connections.

SERIES-CONNECTED RESISTORS

In any series circuit a current I will flow through all parts of the circuit as a result of the potential difference supplied by the battery V_T. Therefore, we say that in a series circuit the current is common throughout that circuit.

When the current flows through each resistor in the circuit, R_1, R_2, and R_3, as, for example, in Fig. 8.5 there will be a voltage drop across that resistor whose value will be determined by the values of I and R, since from Ohm's law $V = I \times R$. The sum of the individual voltage drops, V_1, V_2, and V_3, will be equal to the total voltage V_T

We can summarise these statements as follows.

For any series circuit

I = common throughout the circuit
$V_T = V_1 + V_2 + V_3$

Let us call this Equation 1.

Let us call the total circuit resistance R_T.

From Ohm's law we know that $V = I \times R$ and therefore

total voltage $\qquad V_T = I \times R_T$
voltage drop across R_1 is $\qquad V_1 = I \times R_1$
voltage drop across R_2 is $\qquad V_2 = I \times R_2$
voltage drop across R_3 is $\qquad V_3 = I \times R_3$

Let us call these Equations 2.

We are looking for an expression for the total resistance in any series circuit and, if we substitute Equations 2 into Equation 1 we have:

$$V_T = V_1 + V_2 + V_3$$
$$\therefore I \times R_T = I \times R_1 + I \times R_2 + I \times R_3$$

Now, since I is common to all terms in the equation, we can divide both sides of the equation by I. This will cancel out I to leave us with an expression for the circuit resistance:

$$R_T = R_1 + R_2 + R_3$$

Note The derivation of this formula is given for information only. Craft students need only state the expression: $R_T = R_1 + R_2 + R_3$ for series connections.

PARALLEL-CONNECTED RESISTORS

In any parallel circuit, as shown in Fig. 8.6, the same voltage acts across all branches of the circuit. The total current will divide when it reaches a resistor junction, part of it flowing in each resistor. The sum of the individual currents I_1, I_2 and I_3 for example in Fig. 8.6 will be equal to the total current I_T. We can summarise these statements as follows.

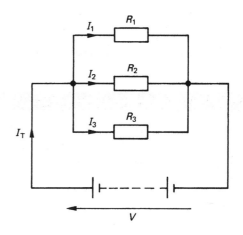

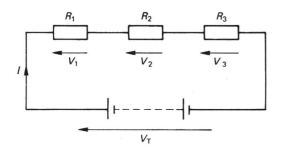

Figure 8.5 A series circuit.

Figure 8.6 A parallel circuit.

For any parallel circuit

V = common to all branches of the circuit
$I_T = I_1 + I_2 + I_3$

Let us call this Equation 1.

Let us call the total resistance R_T.

From Ohm's law we know that $I = \dfrac{V}{R}$ and therefore

total current $\qquad I_T = \dfrac{V}{R_T}$

current through R_1 is $I_1 = \dfrac{V}{R_1}$

current through R_2 is $I_2 = \dfrac{V}{R_2}$

current through R_3 is $I_3 = \dfrac{V}{R_3}$

Let us call these Equations 2.

We are looking for an expression for the equivalent resistance R_T in any parallel circuit and, if we substitute Equations 2 into Equation 1 we have:

$$I_T = I_1 + I_2 + I_3$$

$$\therefore \dfrac{V}{R_T} = \dfrac{V}{R_1} + \dfrac{V}{R_2} + \dfrac{V}{R_3}$$

Now since V is common to all terms in the equation, we can divide both sides by V, leaving us with an expression for the circuit resistance:

$$\dfrac{1}{R_T} = \dfrac{1}{R_1} + \dfrac{1}{R_2} + \dfrac{1}{R_3}$$

Note The derivation of this formula is given for information only. Craft students need only state the expression: $1/R_T = 1/R_1 + 1/R_2 + 1/R_3$ for parallel connections.

EXAMPLE 1

Three $6\,\Omega$ resistors are connected (a) in series, see Fig. 8.7 and (b) in parallel, see Fig. 8.8 across a 12V battery. For each method of connection, find the total resistance and the values of all currents and voltages.

(a) For any series connection

$$R_T = R_1 + R_2 + R_3$$
$$\therefore R_T = 6\,\Omega + 6\,\Omega + 6\,\Omega = 18\,\Omega$$

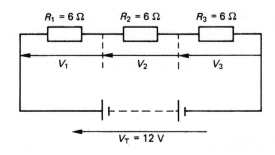

Figure 8.7 Resistors in series.

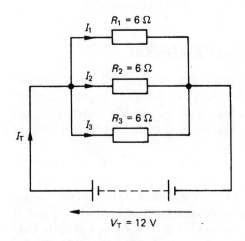

Figure 8.8 Resistors in parallel.

Total current is

$$I_t = \dfrac{V}{R_T}$$

$$\therefore I_t = \dfrac{12\,V}{18\,\Omega} = 0.66\,A$$

The voltage drop across R_1 is

$$V_1 = I \times R_1$$
$$\therefore V_1 = 0.66\,A \times 6\,\Omega = 4\,V$$

The voltage drop across R_2 is

$$V_2 = I \times R_2$$
$$\therefore V_2 = 0.66\,A \times 6\,\Omega = 4\,V$$

The voltage drop across R_3 is

$$V_3 = I \times R_3$$
$$\therefore V_3 = 0.66\,A \times 6\,\Omega = 4\,V$$

For any parallel connection

$$\frac{1}{R_T} = \frac{1}{R_1} + \frac{1}{R_2} + \frac{1}{R_3}$$

$$\therefore \frac{1}{R_T} = \frac{1}{6\,\Omega} + \frac{1}{6\,\Omega} + \frac{1}{6\,\Omega}$$

$$= \frac{1+1+1}{6\,\Omega} = \frac{3}{6\,\Omega}$$

$$\therefore R_T = \frac{6\,\Omega}{3} = 2\,\Omega$$

The total current is

$$I_T = \frac{V}{R_T}$$

$$\therefore I_T = \frac{12\,V}{2\,\Omega} = 6\,A$$

The current flowing through R_1 is

$$I_1 = \frac{V}{R_1}$$

$$\therefore I_1 = \frac{12\,V}{6\,\Omega} = 2\,A$$

The current flowing through R_2 is

$$I_2 = \frac{V}{R_2}$$

$$\therefore I_2 = \frac{12\,V}{6\,\Omega} = 2\,A$$

The current flowing through R_3 is

$$I_3 = \frac{V}{R_3}$$

$$\therefore I_3 = \frac{12\,V}{6\,\Omega} = 2\,A$$

SERIES AND PARALLEL COMBINATIONS

The most complex arrangement of series and parallel resistors can be simplified into a single equivalent resistor by combining the separate rules for series and parallel resistors.

EXAMPLE 1

Resolve the circuit shown in Fig. 8.9 into a single resistor and calculate the potential difference across each resistor. By inspection the circuit contains a parallel group R_3, R_4, R_5, and a series group consisting of R_1 and R_2 in series with the equivalent resistor for the parallel branch. Consider the parallel group R_3, R_4, R_5: we will label this group R_P.

$$\frac{1}{R_P} = \frac{1}{R_3} + \frac{1}{R_4} + \frac{1}{R_5}$$

$$\therefore \frac{1}{R_P} = \frac{1}{2\,\Omega} + \frac{1}{3\,\Omega} + \frac{1}{6\,\Omega}$$

$$= \frac{3+2+1}{6\,\Omega} = \frac{6}{6\,\Omega}$$

$$\therefore R_P = \frac{6}{6\,\Omega} = 1\,\Omega$$

Figure 8.9 may now be represented by the more simple equivalent shown in Fig. 8.10. Since all resistors are now in series

$$R_T = R_1 + R_2 + R_P$$
$$\therefore R_T = 3\,\Omega + 6\,\Omega + 1\,\Omega = 10\,\Omega$$

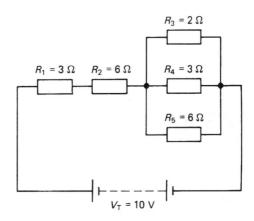

Figure 8.9 A series/parallel circuit.

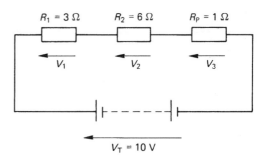

Figure 8.10 Equivalent series circuit.

Thus, the circuit may be represented by a single equivalent resistor of value 10 Ω as shown in Fig. 8.11. The total current flowing in the circuit may be found by using Ohm's law $I = V/R$ as

$$I_T = \frac{V_T}{R_T} = \frac{10\,V}{10\,\Omega} = 1\,A$$

The potential differences across the individual resistors are:

$$V_1 = 1 \times R_1 = 1\,A \times 3\,\Omega = 3\,V$$
$$V_2 = 1 \times R_2 = 1\,A \times 6\,\Omega = 6\,V$$
$$V_P = 1 \times R_P = 1\,A \times 1\,\Omega = 1\,V$$

Since the same voltage acts across all branches of a parallel circuit the same p.d. of one volt will exist across each resistor in the parallel branch R_3, R_4 and R_5. Six volts will be dropped across R_2 and three volts across R_1.

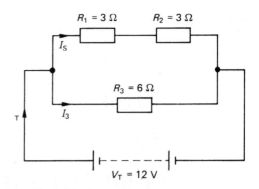

Figure 8.13 Equivalent circuit for Example 2.

Figure 8.13 may now be represented by a more simple equivalent circuit as shown in Fig. 8.14. Since the resistors are now in parallel, the equivalent resistance may be found from

$$\frac{1}{R_T} = \frac{1}{R_s} + \frac{1}{R_3}$$

$$\therefore \frac{1}{R_T} = \frac{1}{6\,\Omega} + \frac{1}{6\,\Omega}$$

$$= \frac{1+1}{6\,\Omega} = \frac{2}{6\,\Omega}$$

$$\therefore R_T = \frac{6\,\Omega}{2} = 3\,\Omega$$

The total current is

$$I_T = \frac{V}{R_T} = \frac{12\,V}{3\,\Omega} = 4\,A$$

Let the current flowing through resistor R_3 be called I_3:

$$\therefore I_3 = \frac{V}{R_3} = \frac{12\,V}{3\,\Omega} = 2\,A$$

Figure 8.11 Single equivalent resistor for Figure 8.9.

EXAMPLE 2

Determine the total resistance and the current flowing through each resistor for the circuit shown in Fig. 8.12. By inspection, it can be seen that R_1 and R_2 are connected in series while R_3 is connected in parallel across R_1 and R_2. The circuit may be more easily understood if we redraw it as in Fig. 8.13. For the series branch, the equivalent resistor can be found from

$$R_s = R_1 + R_2$$
$$\therefore R_s = 3\,\Omega + 3\,\Omega = 6\,\Omega$$

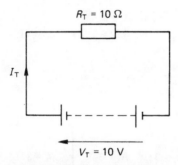

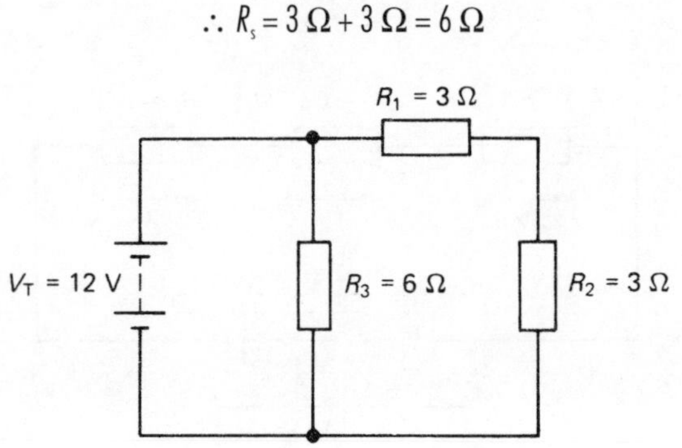

Figure 8.12 A series/parallel circuit for Example 2.

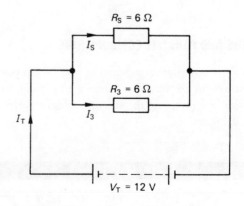

Figure 8.14 Simplified equivalent circuit for Example 2.

Let the current flowing through both resistors R_1 and R_2, as shown in Fig. 8.13, be called I_s:

$$\therefore I_s = \frac{V}{R_s} = \frac{12\,V}{3\,\Omega} = 2\,A$$

CAPACITORS IN COMBINATION

Capacitors, like resistors, may be joined together in various combinations of series or parallel connections, see Fig. 8.15. The equivalent capacitance C_T, of a number of capacitors is found by the application of similar formulae to that used for resistors and discussed earlier in this chapter.

Note The form of the formulae is the opposite way round to that used for series and parallel resistors. The most complex arrangement of capacitors may be simplified into a single equivalent capacitor by applying the separate rules for series or parallel capacitors in a similar way to the simplification of resistive circuits.

EXAMPLE 1

Capacitors of 10 μF and 20 μF are connected first in series and then in parallel, as shown in Figs. 8.16 and 8.17. Calculate the effective capacitance for each connection. Consider the series connection:

$$\frac{1}{C_T} = \frac{1}{C_1} + \frac{1}{C_2}$$

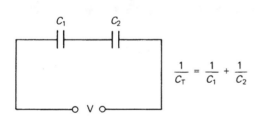

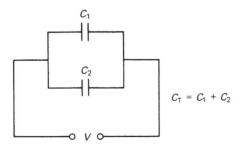

Figure 8.15 Connection and formulae of series or parallel capacitors.

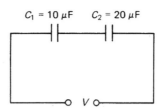

Figure 8.16 Series capacitors.

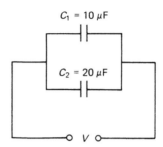

Figure 8.17 Parallel capacitors.

$$\therefore \frac{1}{C_T} = \frac{1}{10\,\mu F} + \frac{1}{20\,\mu F}$$

$$= \frac{2+1}{20\,\mu F} = \frac{3}{20\,\mu F}$$

$$\therefore C_T = \frac{20\,\mu F}{3} = 6.66\,\mu F$$

Consider the parallel connection:

$$C_T = C_1 + C_2$$
$$\therefore C_T = 10\,\mu F + 20\,\mu F = 30\,\mu F$$

Therefore, when capacitors of 10 μF and 20 μF are connected in series their combined effect is equivalent to a single capacitor of 6.66 μF, but when the same capacitors are connected in parallel their combined effect is equal to a capacitor of 30 μF.

Power and Energy

POWER

Power is the rate of doing work and is measured in watts:

$$\text{power} = \frac{\text{work done}}{\text{time taken}}\,(W)$$

In an electrical circuit

$$\text{power} = \text{voltage} \times \text{current}\,(W)$$

Let us call this Equation 1.

Now, from Ohm's law

$$\text{Voltage} = I \times R \text{ (V)}$$

Let us call this Equation 2. Also

$$\text{current} = \frac{V}{R} \text{ (A)}$$

Let us call this Equation 3.
Substituting Equation 2 into Equation 1 we have

$$\text{power} = (I \times R) \times \text{current} = I^2 \times R \text{ (W)}$$

and substituting Equation 3 into Equation 1 we have

$$\text{power} = \text{volts} \times \frac{V}{R} = \frac{V^2}{R} \text{ (W)}$$

We can find the power of a circuit by using any of the three formulae

$$P = V \times I, \quad P = I^2 \times R, \quad P = \frac{V^2}{R}$$

ENERGY

Energy is a concept which engineers and scientists use to describe the ability to do work in a circuit or system.

$$\text{Energy} = \text{Power} \times \text{Time}$$

but, since Power = Voltage × Current then

$$\text{Energy} = \text{Voltage} \times \text{Current} \times \text{Time}$$

The SI unit of energy is the Joule, where time is measured in seconds. For practical electrical installation circuits this unit is very small and, therefore, the kilowatt hour (kWh) is used for domestic and commercial installations. Electricity meters measure 'units' of electrical energy where each 'unit' is one kWh:

energy in joules = voltage × current × time in seconds

and

energy in kWh = kW × time in hours

EXAMPLE 1

A domestic immersion heater is switched on for 40 minutes and takes 13.045 A from a 230 V supply. Calculate the energy used during this time.

Power = Voltage × Current
= 230 V × 13.045 A = 3000 V or 3 kW

Energy = kW × Time in hours

$$= 3 \text{ kW} \times \frac{40 \text{ min}}{60 \text{ min/h}} = 2 \text{ kWh}$$

This immersion heater uses 2 kWh in 40 minutes or 2 'units' of electrical energy every 40 minutes.

EXAMPLE 2

An electronic hi-fi system takes a current of 200 mA when connected to the 230 V mains. Calculate (a) the power, (b) the energy in Joules and (c) the energy in kWh consumed by this system per hour.

(a) Power = Voltage × Current
= 230 V × 200 × 10^{-3} A = 46 W

(b) Energy = Power × Time in seconds
= 46 W × 60 × 60 s = 165 600 J

(c) Energy = kW × Time in hours
= 0.046 × 1 = 0.046 kWh

In general, electronic equipment consumes much less energy than electrical installation appliances. You can also see from the answer to part (b) above that the Joule is a very small unit of energy and, therefore, not a practical unit for most electrical purposes.

EXAMPLE 3

Two 50 Ω resistors may be connected to a 200 V supply. Determine the power dissipated by the resistors when they are connected (a) in series, (b) each resistor separately connected and (c) in parallel.

(a) The equivalent resistance when resistors are connected in series is given by

$$R_T = R_1 + R_2 \text{ Ω}$$
$$\therefore R_T = 50 \text{ Ω} + 50 \text{ Ω} = 100 \text{ Ω}$$
$$\text{Power} = \frac{V^2}{R_T} \text{ (W)}$$
$$\therefore \text{Power} = \frac{200 \text{ V} \times 200 \text{ V}}{100 \text{ Ω}} = 400 \text{ W}$$

(b) Each resistor separately connected has a resistance of 50 Ω:

$$\text{Power} = \frac{V^2}{R} \text{ (W)}$$

$$\therefore \text{Power} = \frac{200 \text{ V} \times 200 \text{ V}}{50 \text{ Ω}} = 800 \text{ W}$$

(c) The equivalent resistance when resistors are connected in parallel is given by

$$\frac{1}{R_T} = \frac{1}{R_1} + \frac{1}{R_2}$$

$$\therefore \frac{1}{R_T} = \frac{1}{50\,\Omega} + \frac{1}{50\,\Omega}$$

$$\frac{1}{R} = \frac{1+1}{50\,\Omega} = \frac{2}{50\,\Omega}$$

$$\therefore R_T = \frac{50\,\Omega}{2} = 25\,\Omega$$

$$\text{Power} = \frac{V^2}{R_T}\ (\text{W})$$

$$\therefore \text{Power} = \frac{200\,V \times 200\,V}{25\,\Omega} = 1600\,W$$

This example shows that by connecting 50 Ω resistors together in different combinations of series of parallel connections, we can obtain various power outputs; in this example 400, 800 and 1600 watts.

INSTRUMENT CONNECTIONS

Electrical and electronic measuring instruments must be chosen and connected into the circuit to be tested with great care for the reasons described in Chapter 6.

Ammeters are connected in series with the load and voltmeters in parallel across the load. Wattmeters contain both current and voltage coils within the instrument. Since watts = volts × amps, the voltage coil must be connected in parallel and the current coil in series with the load as shown in Fig. 8.18.

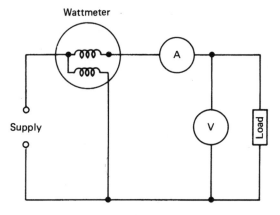

Figure 8.18 Wattmeter, ammeter and voltmeter connected to a load.

Alternating current theory

The supply which we obtain from a car battery is a uni-directional or d.c. supply, whereas the mains electricity supply is alternating or a.c. as shown in Fig. 8.19.

Most electrical equipment makes use of alternating current supplies and for this reason a knowledge of alternating waveforms and their effect upon resistive, capacitive and inductive loads is necessary for all practising electricians.

When a coil of wire is rotated inside a magnetic field a voltage is induced in the coil. The induced voltage follows the mathematical law known as a sinusoidal law and, therefore, we can say that a sine wave has been generated. Such a waveform has the characteristics displayed in Fig. 8.20.

In the UK we generate electricity at a frequency of 50 Hz and the time taken to complete each cycle is given by

$$T = \frac{1}{f}\ (s)$$

$$\therefore T = \frac{1}{50}\ \text{Hz} = 0.02\ s$$

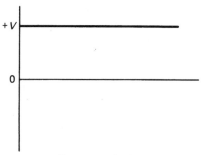

Battery supply d.c.

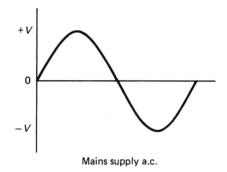

Mains supply a.c.

Figure 8.19 Uni-directional and alternating supply.

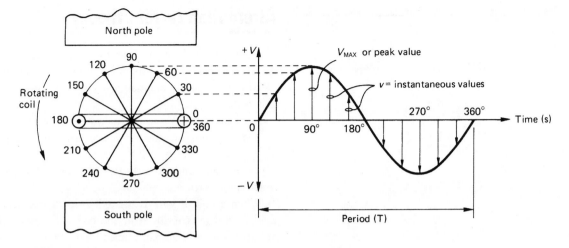

Figure 8.20 Characteristics of a sine wave.

An alternating waveform is constantly changing from zero to a maximum, first in one direction, then in the opposite direction and so the instantaneous values of the generated voltage are always changing. A useful description of the electrical effects of an a.c. waveform can be given by the maximum, average and rms values of the waveform.

The maximum or peak value is the greatest instantaneous value reached by the generated waveform. Cable and equipment insulation levels must be equal to or greater than this value.

The average value is the average over one half-cycle of the instantaneous values as they change from zero to a maximum and can be found from the following formula applied to the sinusoidal waveform shown in Fig. 8.21.

$$V_{AV} = \frac{V_1 + V_2 + V_3 + V_4 + V_5 + V_6}{6}$$

$$= 0.637\ V_{MAX}$$

For any sinusoidal waveform the average value is equal to 0.637 of the maximum value.

The rms value is the square root of the mean of the individual squared values and is the value of an a.c. voltage which produces the same heating effect as a d.c. voltage. The value can be found from the following formulae applied to the sinusoidal waveform shown in Fig. 8.21.

$$V_{rms} = \sqrt{\frac{V_1^2 + V_2^2 + V_3^2 + V_4^2 + V_5^2 + V_6^2}{6}}$$

$$= 0.7071\ V_{MAX}$$

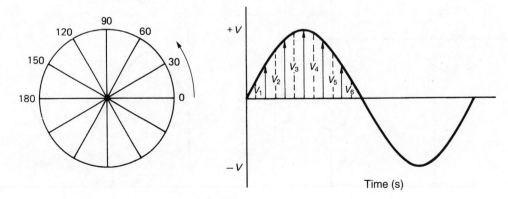

Figure 8.21 Sinusoidal waveform showing instantaneous values of voltage.

For any sinusoidal waveform the rms value is equal to 0.7071 of the maximum value.

EXAMPLE

The sinusoidal waveform applied to a particular circuit has a maximum value of 325.3 V. Calculate the average and rms value of the waveform.

Average value $V_{AV} = 0.637 \times V_{MAX}$
$\therefore V_{AV} = 0.637 \times 325.3 = 207.2$ V

rms value $= V_{rms} = 0.7071 \times V_{MAX}$
$= 0.7071 \times 325.3 = 230$ V

When we say that the main supply to a domestic property is 230 V we really mean 230 V rms. Such a waveform has an average value of about 207 V and a maximum value of about 325 V but because the rms value gives the d.c. equivalent value we almost always give the rms value without identifying it as such.

Resistance and reactance in an a.c. circuit

RESISTANCE (R)

In any d.c. circuit the opposition to current flow is called the resistance of the circuit, measured in ohms and given by the symbol R:

$$R = \frac{V_R}{I_R} \ (\Omega)$$

In an a.c. circuit the total opposition is due to the resistance *and* reactance of the circuit.

INDUCTIVE REACTANCE (X_L)

This is the opposition to an a.c. current in an inductive circuit. It causes the current in the circuit to lag behind the applied voltage as shown in Fig. 8.22. Inductive reactance is given by

$$X_L = \frac{V_L}{I_L} \ (\Omega)$$

or $X_L = 2\pi f L \ (\Omega)$

where 2 = a number constant
π = 3.142 a constant
f = the frequency of the supply
L = the inductance of the circuit

CAPACITIVE REACTANCE (X_c)

This is the opposition to an a.c. current in a capacitive circuit. It causes the current in the circuit to lead

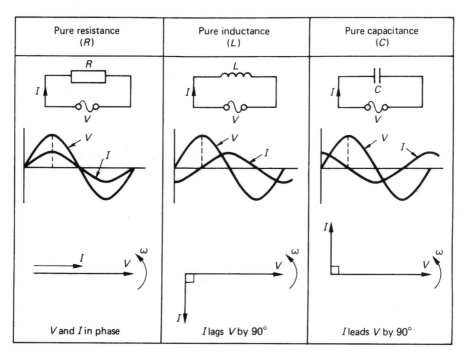

Figure 8.22 Voltage and current relationships in resistive, capacitive and inductive circuits.

ahead of the voltage as shown in Fig. 8.22. Capacitive reactance is given by

$$X_C = \frac{V_C}{I_C} \; (\Omega)$$

$$\text{or } X_C = \frac{1}{2\pi f C} \; (\Omega)$$

where 2 = a number constant
π = 3.142 a constant
f = the frequency of the supply
C = the capacitance of the circuit

When circuits contain two or more separate elements such as RL, RC or RLC, the total opposition to current flow is known as the impedance of the circuit and given the symbol Z.

EXAMPLE 1

Calculate the reactance of a 150 μF capacitor and a 0.05 H inductor if they were separately connected to the 50 Hz mains supply.
Capacitive reactance is given by:

$$X_C = \frac{1}{2\pi f C}$$

where f = 50 Hz and C = 150 μF = 150 × 10⁻⁶ F.

$$\therefore X_C = \frac{1}{2 \times 3.142 \times 50 \text{ Hz} \times 150 \times 10^{-6}}$$

$$= 21.2 \; \Omega$$

Inductive reactance is given by:

$$X_L = 2\pi f L$$

where f = 50 Hz and L = 0.05 H.

$$\therefore X_L = 2 \times 3.142 \times 50 \text{ Hz} \times 0.05 \text{ H} = 15.7 \; \Omega$$

Resistance, inductance and capacitance in an a.c. circuit

When a resistor only is connected to an a.c. circuit the current and voltage waveforms remain together, starting and finishing at the same time. We say that the waveforms are *in phase*.

When a pure inductor is connected to an a.c. circuit the current lags behind the voltage waveform by an angle of 90°. We say that the current *lags* the voltage by 90°. When a pure capacitor is connected to an a.c. circuit the current *leads* the voltage by an angle of 90°. The various effects can be observed on an oscilloscope, but the circuit diagram, waveform diagram and phasor diagram for each circuit are shown in Fig. 8.22.

PHASOR DIAGRAMS

Phasor diagrams and a.c. circuits are an inseparable combination. Phasor diagrams allow us to produce a model or picture of the circuit under consideration, which helps us to understand the circuit. A phasor is a straight line, having definite length and direction, which represents to scale, the magnitude and direction of a quantity such as a current, voltage or impedance.

To find the combined effect of two quantities we combine their phasors by adding the beginning of the second phasor to the end of the first. The combined effect of the two quantities is shown by the resultant phasor, which is measured from the original zero position to the end of the last phasor.

EXAMPLE 1

Find by phasor addition the combined effect of currents A and B acting in a circuit. Current A has a value of 4 A and current B a value of 3 A, leading A by 90°. We usually assume phasors to rotate anti-clockwise and so the complete diagram will be as shown in Fig. 8.23. Choose a scale of, for example, 1 A = 1 cm and draw the phasors to scale, i.e. A = 4 cm and B = 3 cm leading A by 90°.

The magnitude of the resultant phasor can be measured from the phasor diagram and is found to be 5 A acting at a phase angle φ of about 37° leading A. We, therefore, say that the combined effect of currents A and B is a current of 5 A at an angle of 37° leading A.

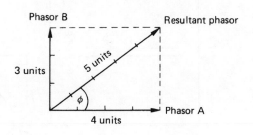

Figure 8.23 The phasor addition of currents A and B.

PHASE ANGLE φ

In an a.c. circuit containing resistance only, such as heating circuits, the voltage and current are in phase, which means that they reach their peak and zero values together, as shown in Fig. 8.24 (a).

In an a.c. circuit containing inductance, such as a motor or discharge lighting circuit, the current often reaches its maximum value after the voltage, which means that the current and voltage are out of phase with each other, as shown in Fig. 8.24 (b). The phase difference, measured in degrees between the current and voltage, is called the phase angle of the circuit, which is denoted by the symbol φ, the small Greek letter phi.

POWER FACTOR

The cosine of the phase angle is known as the power factor (p.f.) of the circuit i.e. cos φ = p.f. If the current leads the voltage we say that power factor is leading, as shown in Fig. 8.24 (c). If the current lags the voltage we say that the power factor is lagging, as shown in Fig. 8.24 (b). This means that we are using the supply voltage as the reference quantity, which seems sensible since the supply authority maintains a constant voltage, but the current depends upon the load. The ideal situation is when V and I are in phase, reaching their maximum and zero values together. The phase angle is then 0° and the power factor is 1 as shown in Fig. 8.24 (a).

The power factor of most industrial loads is lagging because the machines and discharge lighting used in industry are mostly inductive. This causes an additional magnetising current to be drawn from the supply, which does not produce power, but does need to be supplied, making supply cables larger.

EXAMPLE 1

A 250 V supply feeds three 2 kW loads with power factors of 1, 0.8 and 0.4. Calculate the current I at each power factor.

$$I = \frac{P}{V \cos \phi}$$

where P = 2 kW = 2000 W
and V = 250 V

At p.f. = 1

$$I = \frac{2000 \text{ W}}{250 \text{ V} \times 1} = 8 \text{ A}$$

At p.f. = 0.8

$$I = \frac{2000 \text{ W}}{250 \text{ V}} \times 0.8 = 10 \text{ A}$$

At p.f. = 0.4

$$I = \frac{2000 \text{ W}}{250 \text{ V} \times 0.4} = 20 \text{ A}$$

It can be seen from these calculations that a 2 kW load supplied at a power factor of 0.4 would require a 20 A cable, while the same load at unity power factor could be supplied with an 8 A cable. There may also be the problem of higher voltage drops in the supply cables. As a result, the supply authorities encourage installation engineers to improve their power factor to a value close to 1 and sometimes charge penalties if the power factor falls below 0.8.

POWER-FACTOR IMPROVEMENT

Most installations have a low power factor because of the inductive nature of the load. A capacitor has the

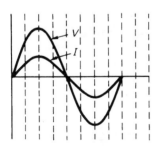

(a) V and I in phase and, therefore, phase angle φ = 0° cos φ = p.f. = 1

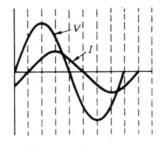

(b) V and I displaced by 45° and, therefore, phase angle φ = 45° cos φ = p.f. = 0.707

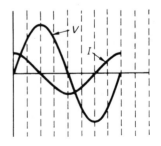

(c) V and I displaced by 90° and, therefore, phase angle φ = 90° cos φ = p.f. = 0

Figure 8.24 Phase relationship of a.c. waveform.

opposite effect to an inductor, and, therefore, we can connect a capacitor in parallel with a load which is known to have a low power factor, in order to improve the power factor of the circuit.

The inductance of the choke in a fluorescent light fitting creates a bad power factor, but by connecting a capacitor across the mains supply to the luminaire, the power factor is improved.

A.C. CIRCUITS

In a circuit containing a resistor and, inductor connected in series, as shown in Fig. 8.25, the current I will flow through the resistor and the inductor, causing the voltages V_R to be dropped across the resistor and V_L to be dropped across the inductor. The sum of these voltages will be equal to the total voltage V_T but as this is an a.c. circuit the voltages must be added by phasor addition. The result is shown in Fig. 8.25 where V_R is drawn to scale and in phase with the current and V_L is drawn to scale and leading the current by 90°. The phasor addition of these two voltages gives us the magnitude and direction of V_T which leads the current by some angle ϕ.

In a circuit containing a resistor and capacitor connected in series, as shown in Fig. 8.26, the current I will flow through the resistor and capacitor, causing voltage drops V_R and V_C. The voltage V_R will be in phase with the current and V_C will lag the current by 90°. The phasor addition of these voltages is equal to the total voltage V_T which, as can be seen in Fig. 8.26, is lagging the current by some angle ϕ.

THE IMPEDANCE TRIANGLE

We have now established the general shape of the phasor diagram for a series a.c. circuit. Figs. 8.25 and 8.26 show the voltage phasors, but we know that $V_R = IR$, $V_L = IX_L$, $V_C = IX_C$ and $V_T = IZ$, and, therefore, the

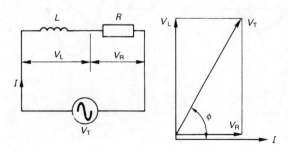

Figure 8.25 A series R-L circuit and phasor diagram.

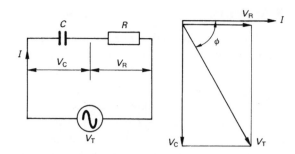

Figure 8.26 A series R-C circuit and phasor diagram.

phasor diagrams (a) and (b) of Fig. 8.27 must be equal. From Fig. 8.27, by the theorem of Pythagoras, we have

$$(IZ)^2 = (IR)^2 + (IX)^2$$
$$I^2 Z^2 = I^2 R^2 + I^2 X^2$$

If we now divide throughout by I^2 we have

$$Z^2 = R^2 + X^2$$
$$\text{or } Z = \sqrt{R^2 + X^2} \; (\Omega)$$

The phasor diagram can be simplified to the impedance triangle given in Fig. 8.27 (c).

EXAMPLE 1

A coil of 0.15 H is connected in series with a 50 Ω resistor across a 100 V 50 Hz supply. Calculate (a) the reactance of the coil, (b) the impedance of the circuit and (c) the current.

(a) $X_L = 2\pi f L \; (\Omega)$
∴ $X_L = 2 \times 3.142 \times 50 \text{ Hz} \times 0.15 \text{ H}$
$= 47.1 \; \Omega$

(b) $Z = \sqrt{R^2 + X^2} \; (\Omega)$
∴ $Z = \sqrt{(50 \;\Omega)^2 + (47.1 \;\Omega)^2} = 68.69 \; \Omega$

(c) $I = V/Z \; (A)$
∴ $I = \dfrac{100 \text{ V}}{68.69 \, \Omega} = 1.46 \text{ A}$

EXAMPLE 2

A 60 μF capacitor is connected in series with a 100 Ω resistor across a 230 V 50 Hz supply. Calculate (a) the reactance of the capacitor, (b) the impedance of the circuit and (c) the current.

(a) $X_C = \dfrac{1}{2\pi f C} \; (\Omega)$

For an inductive circuit

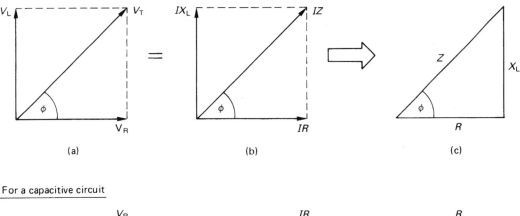

For a capacitive circuit

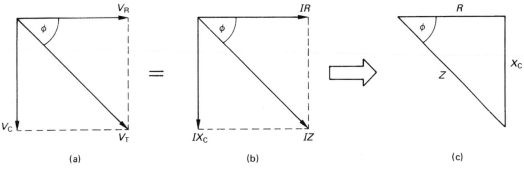

Figure 8.27 Phasor diagram and impedance triangle.

$$\therefore X_c = \frac{1}{2\pi \times 50\text{Hz} \times 60 \times 10^{-6}\,\text{F}}$$
$$= 53.05\,\Omega$$

(b) $Z = \sqrt{R^2 + X^2}\,(\Omega)$
$\therefore Z = \sqrt{(100\,\Omega)^2 + (53.05\,\Omega)^2} = 113.2\,\Omega$

(c) $I = V/Z\,(\text{A})$
$\therefore I = \dfrac{230\,\text{V}}{113.2\,\Omega} = 2.03\,\text{A}$

Power and power factor

A little earlier in this chapter power factor was defined as the cosine of the phase angle between the current and voltage. Power factor may be abbreviated to p.f. If the current lags the voltage, as shown in Fig. 8.25, we say that the p.f. is lagging, and if the current leads the voltage as shown in Fig. 8.26, the p.f. is said to be leading. p.f = cos ϕ.

From the trigonometry of the impedance triangle shown in Fig. 8.27, p.f. is also equal to:

$$\text{p.f.} = \cos\phi = \frac{R}{Z} = \frac{V_R}{V_T}$$

The electrical power in a circuit is the product of the instantaneous value of the voltage and current. Fig. 8.28 shows the voltage and current waveform for a pure indicator and pure capacitor. The power waveform is obtained from the product of V and I at every instant in the cycle. It can be seen that the power waveform reverses every quarter cycle, indicating that energy is alternately being fed into and taken out of the inductor and capacitor. When considered over one complete cycle, the positive and negative portions are equal, showing that the average power consumed by a pure inductor or capacitor is zero. This shows that inductors and capacitors store energy during one part of the voltage cycle and feed it back into the supply later in the cycle. Inductors store

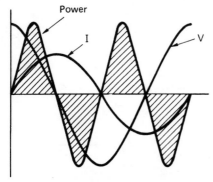

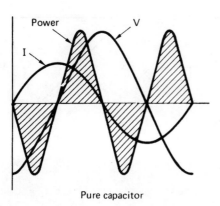

Figure 8.28 Waveform for the a.c. power in purely inductive and purely capacitive circuits.

energy as a magnetic field and capacitors as an electric field.

In an electric circuit more power is taken from the supply than is fed back into it, since some power is dissipated by the resistance of the circuit and, therefore

$$P = I^2 R \text{ (W)}$$

In any d.c. circuit the power consumed is given by the product of the voltage and current, because in a d.c. circuit, voltage and current are in phase. In an a.c. circuit the power consumed is given by the product of the current and that part of the voltage which is in phase with the current. The in-phase component of the voltage is given by $V \cos \phi$ and so power can also be given by the equation

$$P = VI \cos \phi \text{ (W)}$$

EXAMPLE 1

A coil has a resistance of 30 Ω and a reactance of 40 Ω when connected to a 250 V supply. Calculate (a) the impedance, (b) the current, (c) p.f. and (d) the power.

(a) $Z = \sqrt{R^2 + X^2}$ (Ω)
$\therefore Z = \sqrt{(30 \, \Omega)^2 + (40 \, \Omega)^2} = 50 \, \Omega$

(b) $I = V/Z$ (A)
$\therefore I = \dfrac{250 \text{ V}}{50 \, \Omega} = 5 \text{ A}$

(c) p.f. $= \cos \phi = \dfrac{R}{Z}$
$\therefore$ p.f. $= \dfrac{30 \, \Omega}{50 \, \Omega} = 0.6$ lagging

(d) $P = VI \cos \phi$ (W)
$\therefore P = 250 \text{ V} \times 5 \text{ A} \times 0.6 = 750 \text{ W}$

EXAMPLE 2

A capacitor of reactance 12 Ω is connected in series with a 9 Ω resistor across a 150 V supply. Calculate (a) the impedance of the circuit, (b) the current, (c) the p.f. and (d) the power.

(a) $Z = \sqrt{R^2 + X^2}$ (Ω)
$\therefore Z = \sqrt{(9 \, \Omega)^2 + (12 \, \Omega)^2} = 15 \, \Omega$

(b) $I = V/Z$ (A)
$\therefore I = \dfrac{150 \text{ V}}{15 \, \Omega} = 10 \text{ A}$

(c) p.f. $= \cos \phi = \dfrac{R}{Z}$
$\therefore$ p.f. $= \dfrac{9 \, \Omega}{15 \, \Omega} = 0.6$ leading

(d) $P = VI \cos \phi$ (W)
$\therefore P = 150 \text{ V} \times 10 \text{ A} \times 0.6 = 900 \text{ W}$

Resistance, inductance and capacitance in series

The circuit diagram and phasor diagram of an R, L, C series circuit are shown in Fig. 8.29. The voltages across the components are represented by V_R, V_L and

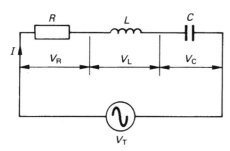

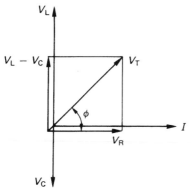

Figure 8.29 R-L-C series circuit and phasor diagram.

V_C, which have the directions shown. Since V_L leads I by 90° and V_C lags by 90° the phasors are in opposition and the combined result is given by $V_L - V_C$ as shown.

Applying the theorem of Pythagoras to the phasor diagram of Fig. 8.29 we have

$$V_T^2 = V_R^2 + (V_L - V_C)^2$$

Since $V_T = IZ$, $V_R = IR$, $V_L = IX_L$ and $V_C = IX_C$ this equation may also be expressed thus:

$$(IZ)^2 = (IR)^2 + (IX_L - IX_C)^2$$

Cancelling out the common factors we have

$$Z^2 = R^2 + (X_L - X_C)^2$$
$$\text{or } Z = \sqrt{R^2 + (X_L - X_C)^2} \ (\Omega)$$

Note The derivation of this equation is not required by craft students but the equation should be remembered and applied in appropriate cases.

EXAMPLE 1

A coil of resistance 5 Ω and inductance 10 mH is connected in series with a 75 μF capacitor across a 200 V 100 Hz supply. Calculate (a) the impedance of the circuit, (b) the current and (c) the p.f.

(a) $X_L = 2\pi f L \ (\Omega)$
∴ $X_L = 2 \times 3.142 \times 100 \text{ Hz} \times 10 \times 10^{-3} \text{ H}$
$= 6.28 \ \Omega$

$X_C = \dfrac{1}{2\pi f C} \ (\Omega)$

∴ $X_C = \dfrac{1}{2 \times 3.142 \times 100 \text{ Hz} \times 75 \times 10^{-6} \text{ F}}$
$= 21.22 \ \Omega$

$Z = \sqrt{R^2 + (X_L - X_C)^2} \ (\Omega)$

∴ $Z = \sqrt{(5 \ \Omega)^2 + (6.28 \ \Omega - 21.22 \ \Omega)^2}$
$= 15.75 \ \Omega$

(b) $I = V/Z \ (A)$

∴ $I = \dfrac{200 \text{ V}}{15.75 \ \Omega} = 12.69 \text{A}$

(c) p.f. $= \cos \phi = \dfrac{R}{Z}$

∴ p.f. $= \dfrac{5}{12.69} = 0.39$

EXAMPLE 2

A 200 μF capacitor is connected in series with a coil of resistance 10 Ω and inductance 100 mH to a 230 V 50 Hz supply. Calculate (a) the impedance, (b) the current and (c) the voltage dropped across each component.

(a) $X_L = 2\pi f L \ (\Omega)$
∴ $X_L = 2 \times 3.142 \times 50 \text{ Hz} \times 100 \times 10^{-3} \text{ H}$
$= 31.42 \ \Omega$

$X_C = \dfrac{1}{2\pi f C} \ (\Omega)$

∴ $X_C = \dfrac{1}{2 \times 3.142 \times 50 \text{Hz} \times 200 \times 10^{-6}}$
$= 15.9 \ \Omega$

$Z = \sqrt{R^2 + (X_L - X_C)^2} \ (\Omega)$

∴ $Z = \sqrt{(10 \ \Omega)^2 + (31.42 \ \Omega - 15.9 \ \Omega)^2}$
$= 18.46 \ \Omega$

(b) $I = V/Z \ (A)$

∴ $I = \dfrac{230 \text{ V}}{18.46 \text{ A}} = 12.46 \text{ A}$

(c) $V_R = I \times R \ (V)$
∴ $V_R = 12.46 \text{ A} \times 10 \ \Omega = 124.6 \text{ V}$

$V_L = I \times X_L \ (V)$
∴ $V_L = 12.46 \text{ A} \times 31.42 \ \Omega = 391.5 \text{ V}$

$V_C = I \times X_C$ (V)

∴ $V_C = 12.46 \text{ A} \times 15.9 \text{ Ω} = 198.1$ V

The phasor diagram of this circuit would be similar to that shown in Fig. 8.29

Series resonance

At resonance the circuit responds sympathetically. Therefore, the condition of resonance is used extensively in electronic and communication circuits for frequency selection and tuning. The current and reactive components of the circuit are at a maximum and so resonance is usually avoided in power applications to prevent cables being overloaded and cable insulation being broken down.

A circuit can be tuned to resonance by either varying the capacitance of the circuit or by adjusting the frequency. At low frequencies the circuit is mainly capacitive and at high frequencies the inductive effect predominates. At some intermediate frequency a point exists where the capacitive effect exactly cancels the inductive effect. This is the point of resonance and occurs when

$$V_L = V_C$$
$$\therefore IX_L = IX_C$$

If we cancel the common factor we have

$$X_L = X_C$$
$$2\pi fL = \frac{1}{2\pi fC}$$

Collecting terms

$$f^2 = \frac{1}{4\pi^2 LC}$$

Taking square roots, resonant frequency is

$$f_0 = \frac{1}{2\pi}\sqrt{\frac{1}{LC}} \text{ (Hz)}$$

Note The resonant frequency is given the symbol f_0. The derivation of the formula is not required by craft students.

At resonance the circuit is purely resistive. $Z = R$, the phase angle is zero and, therefore, the supply voltage and current must be in phase. These effects are shown in Fig. 8.30.

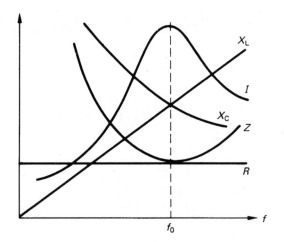

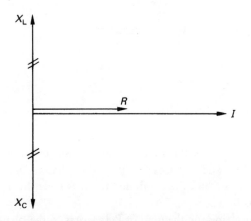

Figure 8.30 Series resonance conditions in an R-L-C circuit.

EXAMPLE 1

A capacitor is connected in series with a coil of resistance 50 Ω and inductance 168.8 mH across a 50 Hz supply. Calculate the value of the capacitor to produce resonance in this circuit.

$X_L = 2\pi fL$ (Ω)

∴ $X_L = 2 \times 3.142 \times 50 \text{ Hz} \times 168.8 \times 10^{-3}$ H
  = 53.03 Ω

At resonance $X_L = X_C$ therefore $X_C = 53.03$ Ω

$$\therefore X_C = \frac{1}{2\pi fC} \text{ (Ω)}$$

Transposing

$$C = \frac{1}{2\pi f X_C} \text{ (F)}$$

$$\therefore C = \frac{1}{2 \times 3.142 \times 50 \text{ Hz} \times 53.03 \text{ Ω}}$$

= 60 μF

EXAMPLE 2

Calculate the resonant frequency f_0 of a circuit consisting of a 25.33 mH inductor connected in series with a 100 μF capacitor.

$$f_0 = \frac{1}{2\pi}\sqrt{\frac{1}{LC}} \text{ (Hz)}$$

$$\therefore f_0 = \frac{1}{2\pi}\sqrt{\frac{1}{25.33 \times 10^{-3}\,\text{H} \times 100 \times 10^{-6}\,\text{F}}}$$

$$= 100 \text{ Hz}$$

Magnetism and motors

Electricity and magnetism have been inseparably connected together since the experiments by Oersted and Faraday in the early nineteenth century. An electric current flowing in a conductor produces a magnetic field *around* the conductor which is proportional to the current. Thus a current of a few milliamperes maintains a weak magnetic field, while lots of current will maintain a strong magnetic field. The magnetic field *spirals* around the conductor as shown in Fig. 8.31 and its direction can be determined by the 'dot' or 'cross notation' and the 'screw rule'. To do this, we imagine the current to be represented by a dart or arrow inside the conductor. The dot represents current coming towards us when we would see the point of the arrow or dart inside the conductor. The cross represents current going away from us when we would see the flights of the dart or arrow.

The magnetic field spirals around the conductor and its direction is determined by the 'screw rule'. That is, we imagine a corkscrew or screw being turned so that it will move in the direction of the current. Therefore, if the current was coming out of the paper as shown in Fig. 8.31 (a) the magnetic field would be spiralling anticlockwise around the conductor. If the current was going into the paper as shown in Fig. 8.31 (b) the magnetic field would spiral clockwise around the conductor.

We know from experiments that lines of magnetic flux never cross and that they behave like stretched elastic bands, always trying to find the shortest straight line between a north and south pole. If we place a current carrying conductor inside a permanent magnetic field as shown in Fig. 8.32 the lines of magnetic flux from the permanent magnet will be distorted to prevent them crossing the lines of magnetic flux produced by the current carrying conductor. Because the distorted lines of magnetic flux belonging to the permanent magnet behave like stretched elastic bands, a force will be exerted on the current carrying conductor, trying to force it out of the permanent magnetic field. This force is the basic working principle of the microphone and loudspeaker (see Fig. 12.18) and of an electric motor, although in a practical electric motor there are many conductors rotating in the magnetic field to ensure sufficient turning force, self starting and smooth running.

Electrical machines are energy converters. If the machine converts electrical energy into mechanical movement, then the machine is a motor. Alternatively, if the machine input is mechanical energy from, say a turbine, and the output is electrical energy, then the machine is a generator, as shown in Fig. 8.33.

A simple generator can be constructed as shown in Fig. 8.34. A simple loop of wire rotated between the poles of a permanent magnet will cut the lines of magnetic flux between the north and south poles. This flux cutting induces an emf or voltage in the loop of wire because of a law discovered by Michael Faraday which states '*when a conductor cuts or is cut by a magnetic field, an emf is induced in that conductor*'. If the generated emf or voltage is collected by carbon brushes at the slip rings and displayed on the screen of a cathode ray oscilloscope, the waveform will be seen to be approximately sinusoidal.

This is the basic working principle of a modern power station generator. In practice the single loop of

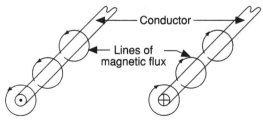

(a) The dot indicates current flowing towards our viewing position

(b) The cross indicates current flowing away from our viewing position.

Figure 8.31 Magnetic fields around a current carrying conductor.

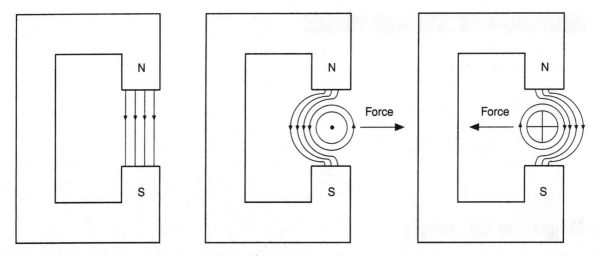

Figure 8.32 Force on a current carrying conductor placed in a magnetic field.

wire would be replaced by many turns, the permanent magnetic field by an electro-magnet and the means of rotating the loops of wire might be by a steam, wind or water turbine, but the generated voltage is produced in exactly the same way.

Transformers

A transformer is an electrical machine which is used to change the value of an alternating voltage. They vary in size from miniature units used in electronics to huge power transformers used in power stations. A transformer will only work when an alternating voltage is connected. It will not normally work from a d.c. supply such as a battery.

A transformer, as shown by the diagram in Fig. 8.35, contains two coils, called the primary and secondary coils, or windings, which are insulated from each other and wound on the same steel or iron core.

An alternating voltage applied to the primary winding produces an alternating current, which sets up an alternating magnetic flux throughout the core. This magnetic flux induces an emf into the secondary winding as described once more by Faraday's Laws, which say '*when a conductor is cut by a magnetic field, an emf is induced in that conductor*'. Since both windings are linked by the same magnetic flux, the induced emf per turn will be the same for both windings. Therefore, the emf in both windings is proportional to the number of turns.

Writing this expression as an equation we have:

the volts per turn on the = the volts per turn
primary winding   on the secondary
                            winding

If we now write this equation as a formula we have:

$$\frac{V_P}{N_P} = \frac{V_S}{N_S} \qquad \text{(Equation 1)}$$

Most practical power transformers have a very high efficiency and for an ideal transformer having a one hundred per cent efficiency

primary power = secondary power
since power = voltage × current then

$$V_P \times I_P = V_S \times I_S \qquad \text{(Equation 2)}$$

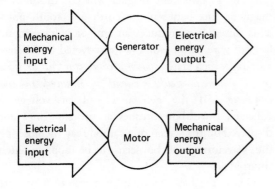

Figure 8.33 Electrical machines as energy converters.

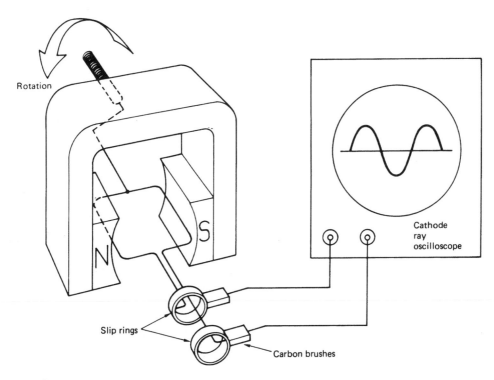

Figure 8.34 Simple a.c. generator.

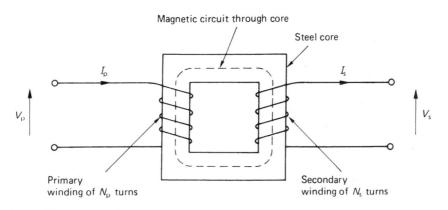

Figure 8.35 A simple transformer.

Combining Equations 1 and 2 we have:

$$\frac{V_P}{V_S} = \frac{N_P}{N_S} = \frac{I_S}{I_P}$$

EXAMPLE 1

A 240 V to 12 V bell transformer is constructed with 800 turns on the primary winding. Calculate the number of secondary turns and the primary and secondary currents when the transformer supplies a 12 V, 12 W alarm bell.

Collecting the information given in the question into a usable form we have:

$V_P = 240$ V
$V_S = 12$ V
$N_P = 800$ turns
Power = 12 W

Information required: N_S, I_S and I_P.

Secondary turns

$$N_S = \frac{N_P V_S}{V_P}$$

$$\therefore N_S = \frac{800 \text{ turns} \times 12 \text{ V}}{240 \text{ V}} = 40 \text{ turns}$$

Secondary current

$$I_S = \frac{\text{Power}}{V_S}$$

$$\therefore I_S = \frac{12 \text{ W}}{12 \text{ V}} = 1 \text{ A}$$

Primary current

$$I_P = \frac{I_S \times V_S}{V_P}$$

$$\therefore I_P = \frac{1 \text{ A} \times 12 \text{ V}}{240 \text{ V}} = 0.05 \text{ A}$$

TRANSFORMER LOSSES

Transformers have a very high efficiency, usually better than 90%, because they have no moving parts causing frictional losses. However, the losses which do occur in a transformer can be grouped under two general headings; copper losses and iron losses.

Copper losses occur because of the small internal resistance of the windings. They are proportional to the load, increasing as the load increases because copper loss is an '$I^2 R$' loss.

Iron losses are made up of *hysteresis loss* and *eddy current loss*. The hysteresis loss depends upon the type of iron used to construct the core and consequently core materials are carefully chosen. Transformers will only operate on an alternating current, thus the current which establishes the core flux is constantly changing from positive to negative. Each time there is a current reversal, the magnetic flux reverses and it is this building up and collapse of magnetic flux in the core material which accounts for the hysteresis loss.

Eddy currents are circulating currents created in the core material by the changing magnetic flux.

The iron loss is a constant loss consuming the same power from no load to full load.

TRANSFORMER CONSTRUCTION

Transformers are constructed in a way which reduces the losses to a minimum. The core of a power transformer is usually made of silicon-iron laminations, because at fixed low frequencies silicon-iron has a small hysteresis loss and the laminations reduce the eddy current loss.

At radio frequencies and above, steel cores cannot be used because the losses become excessive. One solution has been to form *ferrite* cores for transformers used in electronic circuits. Ferrite cores are made by suspending iron dust in a non-conducting bakelite medium.

The primary and secondary windings are wound close to each other on the central limb of the transformer core. If the windings are spread over two limbs, there will usually be half of each winding on each limb as shown in Fig. 8.36.

MATCHING TRANSFORMERS

Transformers are sometimes used as an impedance matching device between a load of low impedance, such as an audio speaker, and an amplifier of high output impedance. The purpose of the transformer is

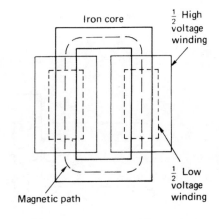
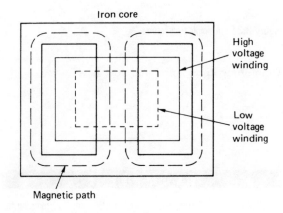

Figure 8.36 Transformer construction.

to make the load presented to the output terminals of the amplifier '*look*' as though it has the same impedance as the amplifier because this ensures maximum power transfer between the amplifier and the load.

This arrangement is shown in Fig. 8.37, where

- Z_P is the amplifier output presented to the primary terminals of the transformer
- Z_S is the load impedance connected to the secondary terminals of the transformer
- N_P is the number of primary turns
- N_S is the number of secondary turns

For a matching transformer

$$\frac{N_P}{N_S} = \sqrt{\frac{Z_P}{Z_S}}$$

EXAMPLE 1

The output impedance of an audio amplifier is 800 Ω and it is proposed to connect an 8 Ω speaker to this amplifier. Determine the turns ratio of the transformer which would ensure maximum power transfer between the amplifier and the speaker.

$$\frac{N_P}{N_S} = \sqrt{\frac{Z_P}{Z_S}}$$

$$\therefore \frac{N_P}{N_S} = \sqrt{\frac{800\ \Omega}{8\ \Omega}} = 10$$

Therefore, a transformer having a turns ratio of 10:1 is required.

EXAMPLE 2

A transformer with a turns ratio of 20:1 is to be used to connect a speaker to a 2 kΩ amplifier. Determine the value of the speaker which will produce maximum power transfer.

$$\frac{N_P}{N_S} = \sqrt{\frac{Z_P}{Z_S}}$$

$$\therefore \frac{20}{1} = \sqrt{\frac{2\ k\Omega}{Z_S}}$$

$$\frac{20^2}{1} = \frac{2\ k\Omega}{Z_S}$$

$$Z_S = \frac{2\ k\Omega}{20^2}$$

$$\therefore Z_S = \frac{2000\ \Omega}{400}$$

$$= 5\ \Omega$$

A 5 Ω speaker will ensure maximum power is transferred from the amplifier.

Mechanics

Mechanics is the scientific study of how things behave when they are acted upon by pushing and pulling forces. Some of the most famous experiments were

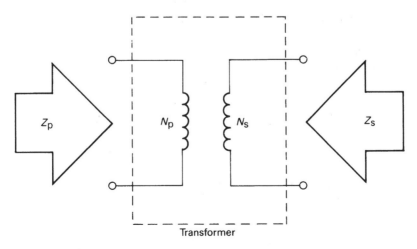

Figure 8.37 Matching transformer connections.

performed by Galileo Galilei in Italy around 1610 and by Sir Isaac Newton in England around 1670, but let us begin by defining some of the terms.

MASS

This is a measure of the amount of material in a substance such as metal, plastic, wood, brick or tissue which is collectively known as a body. The SI unit of mass is the kilogram abbreviated to kg.

SPEED

The feeling of speed is something with which we are all familiar. If we travel in a motor vehicle we know that an increase in speed would, excluding accidents, allow us to arrive at our destination more quickly. Therefore, speed is concerned with distance travelled and time taken. Suppose we were to travel a distance of 30 miles in one hour, our speed would be an average of 30 miles per hour:

$$\text{Speed} = \frac{\text{Distance (m)}}{\text{Time (s)}}$$

VELOCITY

In everyday conversation we often use the word velocity to mean the same as speed, and indeed the units are the same. However, for scientific purposes this is not acceptable since velocity is also concerned with direction. Velocity is speed in a given direction. For example, the speed of an aircraft might be 200 miles per hour, but its velocity would be 200 miles per hour in, say a westerly direction. Speed is a scalar quantity where velocity is a vector quantity.

$$\text{Velocity} = \frac{\text{Distance (m)}}{\text{Time (s)}}$$

ACCELERATION

When an aircraft takes off, it starts from rest and increases its velocity until it can fly. This change in velocity is called its acceleration. By definition acceleration is the rate of change in velocity with time:

$$\text{Acceleration} = \frac{\text{Velocity}}{\text{Time}} \ (m/s^2)$$

The SI unit for acceleration is the metre per second per second or m/s^2.

EXAMPLE 1

If an aircraft accelerates from a velocity of 15 m/s to 35 m/s in 4 s calculate its average acceleration:

$$\text{Average velocity} = 35 \text{ m/s} - 15 \text{ m/s} = 20 \text{ m/s}$$

$$\text{Average acceleration} = \frac{\text{Velocity}}{\text{Time}} = 20/4 = 5 \text{ m/s}^2$$

Thus, the average acceleration is 5 metres per second, every second.

FORCE

The presence of a force can only be detected by its effect on a body. A force may cause a stationary object to move or bring a moving body to rest. For example, a number of people pushing a broken down motor car exert a force which propels it forward, but applying the motor car brakes applies a force on the brake drums which slows down or stops the vehicle. Gravitational force causes objects to fall to the ground. The apple fell from the tree on to Isaac Newton's head as a result of gravitational force. The standard rate of acceleration due to gravity is accepted as 9.81 m/s^2. Therefore, an apple weighing 1 kg will exert a force of 9.81 N since

$$\text{Force} = \text{Mass} \times \text{Acceleration (N)}$$

The SI unit of force is the newton, symbol N, to commemorate the great English scientist Sir Isaac Newton (1642–1727).

EXAMPLE

A 50 kg bag of cement falls from a forklift truck whilst being lifted to a storage shelf. Determine the force with which the bag will strike the ground.

$$\text{Force} = \text{Mass} \times \text{Acceleration (N)}$$
$$\therefore \text{Force} = 50 \text{ kg} \times 9.81 \text{ m/s}^2 = 490.5 \text{ N}$$

PRESSURE OR STRESS

To move a broken motor car I might exert a force on the back of the car to propel it forward. My hands would apply a pressure on the body panel at the point of contact with the car. Pressure or stress is a measure of the force per unit area.

$$\text{Pressure or stress} = \frac{\text{Force}}{\text{Area}} \ (N/m^2)$$

EXAMPLE 1

A young woman of mass 60 kg puts all her weight on to the heel of one shoe which has an area of 1 cm². Calculate the pressure exerted by the shoe on the floor. (Assume the acceleration due to gravity to be 9.81 m/s².)

$$\text{Pressure} = \frac{\text{Force}}{\text{Area}} \; (N/m^2)$$

$$\therefore \text{Pressure} = \frac{60 \text{ kg} \times 9.81 \text{ m/s}^2}{1 \times 10^{-4} \text{ m}^2} = 5886 \text{ kN/m}^2$$

EXAMPLE 2

A small circus elephant of 1 tonne (1000 kg) puts all its weight on to one foot which has a surface area of 400 cm². Calculate the pressure exerted by the elephant's foot on the floor, assuming the acceleration due to gravity to be 9.81 m/s².

$$\text{Pressure} = \frac{\text{Force}}{\text{Area}} \; (N/m^2)$$

$$\therefore \text{Pressure} = \frac{1000 \text{ kg} \times 9.81 \text{ m/s}^2}{400 \times 10^{-4}} = 245.3 \text{ kN/m}^2$$

These two examples show that the young woman exerts 24 times more pressure on the ground than the elephant. This is because her mass exerts a force over a much smaller area than the elephant's foot, and is the reason why many wooden floors are damaged by high heeled shoes.

WORK DONE

Suppose a broken motor car was to be pushed along a road, work would be done on the car by applying the force necessary to move it along the road. Heavy breathing and perspiration would be evidence of the work done.

Work done = Force × Distance moved in the direction of the force (J)

The SI unit of work done is the Newton metre or Joule. The Joule is the preferred unit and it commemorates an English physicist, James Prescot Joule (1818–1889).

EXAMPLE

A building hoist lifts ten 50 kg bags of cement through a vertical distance of 30 m to the top of a high rise building. Calculate the work done by the hoist, assuming the acceleration due to gravity to be 9.81 m/s².

Work done = Force × Distance moved (J)

But Force = Mass × Acceleration (N). By substitution:

Work done = Mass × Acceleration × Distance moved
$$\therefore \text{Work done} = 10 \times 50 \text{ kg} \times 9.81 \text{ m/s}^2 \times 30 \text{ m}$$
$$= 147.15 \text{ kJ}$$

POWER

If a motor car can cover the distance between two points more quickly than another car, we say that the faster car is more powerful. It can do a given amount of work more quickly. By definition, power is the rate of doing work.

$$\text{Power} = \frac{\text{Work done}}{\text{Time taken}} \; (W)$$

The SI unit of power, both electrical and mechanical, is the watt symbol W. This commemorates the names of James Watt (1736–1819) the inventor of the steam engine.

EXAMPLE 1

A building hoist lifts ten 50 kg bags of cement to the top of a 30 m high building. Calculate the rating (power) of the motor to perform this task in 60 seconds if the acceleration due to gravity is taken as 9.81 m/s².

$$\text{Power} = \frac{\text{Work done}}{\text{Time taken}} \; (W)$$

But Work done = Force × Distance moved (J) and Force = Mass × Acceleration (N). By substitution

$$\text{Power} = \frac{\text{Mass} \times \text{Acceleration} \times \text{Distance moved}}{\text{Time taken}} \; (W)$$

$$\therefore \text{Power} = \frac{10 \times 5 \text{ kg} \times 9.81 \text{ m/s}^2 \times 30 \text{ m}}{60 \text{ s}}$$

$$= 2452.5 \text{ W}$$

The rating of the building hoist motor will be 2.45 kW.

EXAMPLE 2

A hydro-electric power station pump motor working continuously during a seven hour period raises 856 tonne of water through a vertical distance of 60 m. Determine the rating (power) of the motor, assuming the acceleration due to gravity is 9.81 m/s².

From Example 1

$$\text{Power} = \frac{\text{Mass} \times \text{Acceleration} \times \text{Distance moved}}{\text{Time taken}} \text{ (W)}$$

$$\therefore \text{Power} = \frac{856 \times 100 \text{ kg} \times 9.81 \text{ m/s}^2 \times 60 \text{ m}}{7 \times 60 \times 60 \text{ s}}$$

$$= 20\,000 \text{ W}$$

The rating of the pump motor is 20 kW.

ENERGY

Have you ever observed how some people seem to have more energy than others? Some footballers appear to be able to run around the sports field all day and remain apparently fresh, whilst others are exhausted after a short dash. The player with most energy is not only a powerful player but can persist in his efforts for a longer period of time.

$$\text{Energy} = \text{Power} \times \text{Time (J)}$$

Anything which is capable of doing work is said to possess energy and so the SI unit of energy is the same as for work; the joule, symbol J.

Energy in the scientific meaning can take many forms, electrical, mechanical, thermal, chemical and atomic, but in mechanics, energy is divided into two types; potential and kinetic energy.

Potential energy is the energy a body possesses because of its position or state of compression. A body raised some distance over the ground possesses potential energy since its mass can do work as it returns to the ground. A wound up spring possesses potential energy because, for example a clock spring does work against the forces opposing the motion of the clock mechanism.

Kinetic energy is the energy a body possesses because of its motion. The water which flows from the dam of a hydro-electric power station possesses kinetic energy since the rapidly flowing water does work by turning a turbine and generating electricity. A hammer head possesses kinetic energy since it can do work by striking a nail and driving the nail into a piece of wood.

Conservation of energy

As you will be aware, the efficiency of a machine is never 100%. In converting energy from one form to another the useful output is always less than the input. This does not mean that energy is destroyed, only that some of the useful energy is lost in the conversion. In the case of an electrical to mechanical conversion, some of the input energy is changed to heat in overcoming friction.

In all energy conversions some of the useful energy is lost but it is not destroyed. It is merely converted to a form which, in most cases, is undesirable or irretrievable. The Conservation of Energy Theory states that '*energy can neither be created nor destroyed, but can be transferred from one form to another*'.

Temperature and heat

Heat has the capacity to do work and is a form of energy. Temperature is not an energy unit but describes the 'hotness' of a substance or material.

Heat transfer

Heat energy is transferred by three separate processes which can occur individually or in combination. The processes are convection, radiation and conduction.

CONVECTION

Air which passes over a heated surface expands, becomes lighter and warmer and rises, being replaced by descending cooler air. These circulating currents of air are called convection currents. In circulating, the warm air gives up some of its heat to the surfaces over which it passes and so warms a room and its contents.

RADIATION

Molecules on a metal surface vibrating with thermal energies generate electromagnetic waves. The waves travel away from the surface at the speed of light taking energy with them and leaving the surface cooler. If the waves meet another material they produce a disturbance of the surface molecules which raises the temperature of that material. Radiated heat requires no intervening medium between the transmitter and receiver, obeys the same laws as light energy and is the method by which the energy from the sun reaches the earth.

CONDUCTION

Heat transfer through a material by conduction occurs because there is direct contact between the vibrating molecules of the material. The application of a heat source to the end of a metal bar causes the atoms to vibrate rapidly within the lattice framework of the material. This violent shaking causes adjacent atoms to vibrate and liberates any loosely bound electrons, which also pass on the heat energy. Thus the heat energy travels **through** the material by conduction.

HEAT DISSIPATION

In order to continue to operate efficiently many electronic components such as transistors, thyristors and triacs need to be able to dissipate or remove the heat generated during normal operation. This is usually achieved by securing the electronic component on to a 'heat sink' which is a large metal surface. Heat then travels *from* the electronic component during normal operation into the metal heat sink by *conduction*. The heat sink then gives up its heat to the surroundings by *convection*. The most efficient shape for a heat sink is one which has a large surface area. Commercial heat sinks are usually made from cast aluminium, which is a good conductor of heat, and often contain many cooling fins to increase the surface area.

In most equipment natural convection cooling is adequate but in exceptional circumstances air can be forced over the heat sink with a fan in order to increase the cooling rate.

EFFECTS OF TEMPERATURE CHANGE

Since heat is a form of energy, a change in temperature will cause a change in a solid, liquid or gas which may or may not be apparent to an observer. To understand these changes we have to understand a little of the composition of materials.

All materials are made up of tiny particles. In a solid material the particles are closely packed together and held very rigidly by strong bonding forces. The bonding forces behave like springs or elastic bands holding the particles together. If the solid is heated, the energy going into it causes the particles to vibrate against the bonding forces and the solid material expands. Railway lines have gaps in them to prevent the rails from buckling as they expand in the summer. Different solids expand at different rates; the bi-metal strips used in thermostats use the different expansion rates of two metals fused together to switch an electrical contact.

If a metal is heated to a very high temperature the bonds are temporarily broken and the material becomes plastic or bendable and can be re-shaped. Upon cooling the bonds re-form and the material becomes solid once more.

A liquid is also made up of particles held together by bonding forces. However, because a liquid can flow and pour, the particles must have a greater freedom to move and, therefore, the bonding forces in a liquid must be less than those in a solid material.

If a liquid is heated, the particles vibrate against the bonding forces and, just like a solid, the liquid expands. This principle is used to practical effect in thermometers and automatic sprinkler systems. The water valve in a sprinkler is held in place by a small glass container filled with liquid. If the surrounding temperature rises above a predetermined level, the liquid expansion shatters the fragile glass container allowing the valve to open and spray the area with water.

The density of a gas is about one thousand times less than the density of a liquid or solid. This suggests that gas particles are much farther apart than solid or liquid particles and that the bonding forces between individual particles is very weak.

At normal temperatures, gas particles move about within a gas container quite independently, occasionally colliding with another particle and the walls of the container. Each collision exerts a force or pressure on the container. If the temperature of a gas is increased the energy which goes into it causes the individual particles to move about more energetically, the collisions within the container become more violent and the pressure increases.

The internal combustion engine uses high temperature gases to drive our motor vehicles. Ignited petrol vapour forces the piston down the cylinder to turn the crankshaft which is connected through the gearbox to the road wheels.

Sound

Sound is by definition a pressure wave in a medium which is audible to the human ear. The human ear can only detect sound waves within the range 15 Hz to about 15 kHz. Frequencies above this range are audible to dogs and bats and below this range to whales, but they are not audible to humans.

Sound waves travel through a medium by creating pressure wave fronts within the medium. It is, therefore, impossible for sound to travel through a vacuum. If a tuning fork is vibrated as shown in Fig. 8.38 it alternately compresses and rarefies the air particles next to the fork. Every compression and rarefaction spreads outwards from the tuning fork and when this is picked up by the human ear we recognise it as a note or sound.

A tuning fork produces a single frequency note which is dependent upon the number of tuning fork vibrations per second. The musical notes doh, ray, me can be produced by tuning forks vibrating at 256, 288 and 320 vibrations per second respectively. A piano produces its wide range of musical notes by vibrating strings in this way.

Sound travels through different mediums at different speeds or velocities, in air, sound travels at about 330 m/s, in water 1450 m/s and in steel sound travels at 5000 m/s.

As the sound wave moves through the air, the distance between the wave fronts is constant. The distance between two successive wave fronts is called the wavelength and is measured in metres. The wave also has a frequency of vibration determined by the frequency of the vibrating tuning fork or instrument string. Velocity, wavelength and frequency are all related by the following equation:

$$v = f\lambda \text{ m/s}$$

where v = the velocity of the sound wave measured in m/s

f = the frequency measured in Hz

λ = the wavelength measured in metres (λ is the Greek letter lambda)

EXAMPLE 1

Determine the wavelength of a sound wave emitted from a speaker at a frequency of 100 Hz if it is travelling in air at a velocity of 33 m/s.

From the formula $v = f\lambda$ (m/s) we can transpose for $\lambda = v/f$ (m)

$$\therefore \lambda = \frac{330 \text{ m/s}}{100 \text{ Hz}} = 3.3 \text{ metre}$$

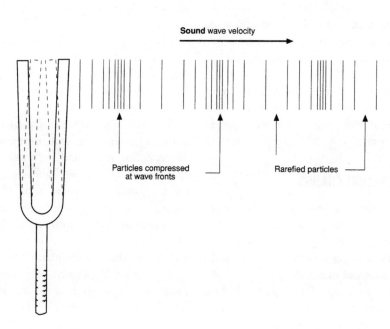

Figure 8.38 Sound waves from a tuning fork.

EXAMPLE 2

Determine the wavelength of another sound wave emitted from a speaker at 10 kHz if it travels in air at 330 m/s.

$$\lambda = \frac{v}{f} \text{ (m)}$$

$$\therefore \lambda = \frac{330s}{10\ 000\ Hz} = 33 \times 10^{-3} \text{ (m) or 33 (mm)}$$

The sound emitted from the speaker at 100 Hz has a wavelength of over a metre and we would recognise the sound as a low base note. The 10 kHz note is toward the upper end of our hearing range, has a much smaller wavelength and would appear to us as a high note. Bands and orchestras use a very wide range of frequencies to make music and modern hi-fi equipment tends to use large woofer speakers for the low frequency base notes which have a long wavelength and small tweeter speakers for the high frequency notes which have a small wavelength, so that a faithful reproduction of the music can be brought into our homes.

Light

Light is by definition an electromagnetic wave which has the correct frequencies required to stimulate the nerve endings in the human eye. Light is only a very small part of the whole electromagnetic spectrum, being made up of a range of frequencies with a very small wavelength in the region of about 10^{-7} metre, that is a wavelength which is smaller than a single grain of talcum powder. Different wavelengths of light are detected by our eyes as different colours. Red has a wavelength of 6.6×10^{-7} m, then come orange, yellow, green, blue, indigo and finally violet, which has a wavelength of 3.3×10^{-7} m.

Sunlight or white light is made up of those seven colours combined. This was *one* of the famous discoveries made by the English scientist Sir Isaac Newton in 1666 when he passed a beam of sunlight through a prism and observed the colours of the rainbow as shown in Fig. 8.39. Then, using a second prism he was able to combine the colours together again to form white light which proved that white light is made up of a combination of colours.

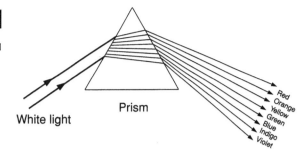

Figure 8.39 White light being refracted into its seven constituent parts.

The scientific definitions and formulae which we applied previously to sound waves also apply to light waves even though the frequency and velocity of visible light is much higher than sound. Light travels at 3×10^8 m/s or 186 000 miles per second, sound only travels at 330 m/s or 720 miles per hour. This is one of the reasons why we see the lightning before we hear the thunder.

EXAMPLE 1

Determine the frequency of red light whose wavelength is 6.6×10^{-7} m. Assume the velocity of light to be 3×10^8 m/s.

From the standard formula $v = f\lambda$ (m/s) and transposing for frequency we have

$$f = \frac{v}{\lambda} \text{ Hz}$$

$$\therefore f = \frac{3 \times 10^8 \text{ m/s}}{6.6 \times 10^{-7} \text{ m}} = 4.545 \times 10^{14} \text{ Hz}$$

EXAMPLE 2

Determine the frequency of violet light whose wavelength is 3.3×10^{-7} m. Assume the velocity of light to be 3×10^8 m/s.

As previously

$$f = \frac{v}{\lambda} \text{ Hz}$$

$$\therefore f = \frac{3 \times 10^8 \text{ ms}}{3.3 \times 10^{-7} \text{ m}} = 9.09 \times 10^{14} \text{ Hz}$$

You can see from these examples that although the wavelengths of the visible spectrum are extremely small and the frequencies incomprehensibly large

(10^{14} Hz = 100 000 000 000 000 Hz) our very sensitive eyes are able to detect and differentiate between the wide range of colours around us.

Reflection and refraction

Light waves travel in straight lines, a feature which scientists call rectilinear propagation which literally means straight line transmission. Light rays can be *reflected* or bounced off a bright surface. The direction of the reflected ray will be determined by the shape of the reflector and the angle at which the light ray strikes the reflector.

If light passes *through* a material the light ray is slowed down and bent away from its original path as it passes through the material. It is this bending of a light ray as it passes through a material which is known as *refraction* and is demonstrated by the light rays passing through the prism shown in Fig. 8.39. Sunlight shining through water droplets is refracted and results in the phenomenon we recognise as a rainbow.

Lenses

A lens is a piece of transparent material which refracts light rays. It is especially manufactured to have the properties of either converging or diverging the light rays as shown in Fig. 8.40.

A converging lens brings all the light rays to a focus at the focal point. A diverging lens spreads out the light rays so that they appear to have come from a focal point behind the lens.

There are many practical applications for thin lenses. Spectacles are used to focus an image at the correct focal length on the back of the eye so that the image is clear and sharp. The camera lens focuses an image on the film and a projector's lens is used to focus an image on a screen.

Colour

Light and colour are a visual sensation. The eye 'sees' colour because it recognises the impact of light waves of different frequencies which make up the visible spectrum. As we said earlier, the visible spectrum is made up of seven colours, but good colour reproduction can be achieved by using only three of these colours, that is, red, blue and green. These are called the primary colours and other colours, secondary colours, can be produced by combining these colours as shown in Fig. 8.41. Magenta, for example can be produced by combining red and blue light, yellow by combining green and red.

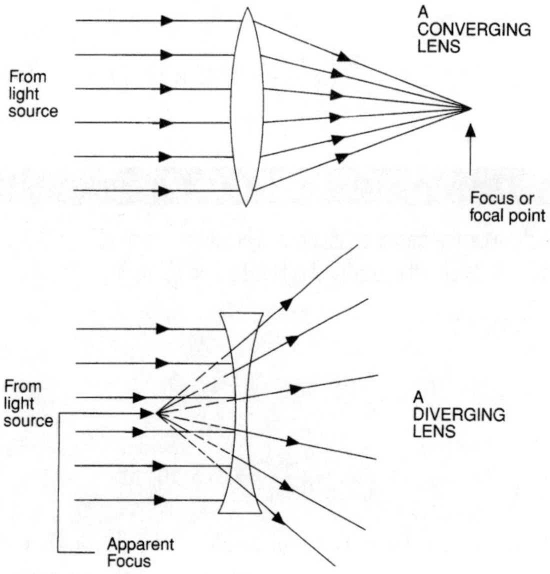

Figure 8.40 Effect of a thin lens on light rays.

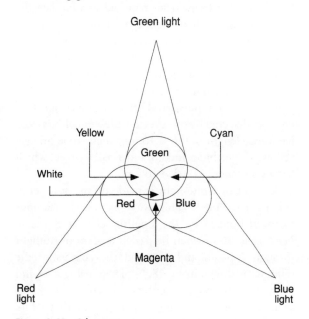

Figure 8.41 Colour mixing.

In a television studio, the television camera views the scene through a red, blue and green filter so that three electronic signals are produced and transmitted. The colour television set at home uses these three signals to cause dots of red, blue and green light to appear on the screen. Our eyes then see a coloured image which is a faithful reproduction of the scene in the television studio.

Exercises

1. The SI units of length, resistance and power are: (a) millimetre, ohm, kilowatt (b) centimetre, ohm, watt (c) metre, ohm, watt (d) kilometre, ohm, kilowatt.
2. The current taken by a 10 Ω resistor when connected to a 240 V supply is: (a) 41 mA (b) 2.4 A (c) 24 A (d) 240 A.
3. The resistance of an element which takes 12 A from a 240 V supply is: (a) 2.88 Ω (b) 5 Ω (c) 12.24 Ω (d) 20 Ω.
4. A 12 Ω lamp was found to be taking a current of 2 A at full brilliance. The voltage across the lamp under these conditions was: (a) 6 V (b) 12 V (c) 24 V (d) 240 V.
5. The resistance of 100 m of 1 mm² cross-section copper cable of resistivity 17.5×10^{-9} Ωm will be: (a) 1.75 mΩ (b) 1.75 Ω (c) 17.5 Ω (d) 17.5 kΩ.
6. The resistance of a motor field winding at 0°C was found to be 120 Ω. Find its new resistance at 20°C if the temperature co-efficient of the winding is 0.004 Ω/Ω°C. (a) 116.08 Ω (b) 120.004 Ω (c) 121.08 Ω (d) 140.004 Ω.
7. The resistance of a motor field winding was found to be 120 Ω at an ambient temperature of 20°C. If the temperature co-efficient of resistance is 0.004 Ω/Ω°C the resistance of the winding of 60°C will be approximately: (a) 102 Ω (b) 120 Ω (c) 130 Ω (d) 138 Ω.
8. A capacitor is charged by a steady current of 5 mA for 10 s. The total charge stored on the capacitor will be: (a) 5 mC (b) 50 mC (c) 5 C (d) 50 C.
9. When 100 V was connected to a 20 µF capacitor the charge stored was: (a) 2 mC (b) 5 mC (c) 20 mC (d) 100 mC.
10. Resistors of 6 Ω and 3 Ω are connected in series. The combined resistance value will be: (a) 2 Ω (b) 3.6 Ω (c) 6.3 Ω (d) 9 Ω.
11. Resistors of 3 Ω and 6 Ω are connected in parallel. The equivalent resistance will be: (a) 2 Ω (b) 3.6 Ω (c) 6.3 Ω (d) 9 Ω.
12. Three resistors of 24 Ω, 40 Ω and 60 Ω are connected in series. The total resistance will be: (a) 12 Ω (b) 26.4 Ω (c) 44 Ω (d) 124 Ω.
13. Resistors of 24 Ω, 40 Ω and 60 Ω are connected together in parallel. The effective resistance of this combination will be: (a) 12 Ω (b) 26.4 Ω (c) 44 Ω (d) 124 Ω.
14. Two identical resistors are connected in series across a 12 V battery. The voltage drop across each resistor will be: (a) 2 V (b) 3 V (c) 6 V (d) 12 V.
15. Two identical resistors are connected in parallel across a 24 V battery. The volt drop across each resistor will be: (a) 6 V (b) 12 V (c) 24 V (d) 48 V.
16. A 6 Ω resistor is connected in series with a 12 Ω resistor across a 36 V supply. The current flowing through the 6 Ω resistor will be: (a) 2 A (b) 3 A (c) 6 A (d) 9 A.
17. A 6 Ω resistor is connected in parallel with a 12 Ω resistor across a 36 V supply. The current flowing through the 12 Ω resistor will be: (a) 2 A (b) 3 A (c) 6 A (d) 9 A.
18. Capacitors of 24 µF, 40 µF and 60 µF are connected in series. The equivalent capacitance will be: (a) 12 µF (b) 44 µF (c) 76 µF (d) 124 µF.
19. Capacitors of 24 µF, 40 µF and 60 µF are connected in parallel. The total capacitance will be: (a) 12 µF (b) 44 µF (c) 76 µF (d) 124 µF.
20. The total power dissipated by a 6 Ω and 12 Ω resistor connected in parallel across a 36 V supply will be: (a) 72 W (b) 324 W (c) 576 W (d) 648 W.
21. Three resistors are connected in series and a current of 10 A flows when they are connected to a 100 V supply. If another resistor of 10 Ω is connected in series with the three series resistors, the current carried by this resistor will be: (a) 4 A (b) 5 A (c) 10 A (d) 100 A.
22. The rms value of a sinusoidal waveform whose maximum value is 100 V will be: (a) 63.7 V (b) 70.71 V (c) 100 V (d) 100.67 V.
23. The average value of a sinusoidal alternating current whose maximum value is 10 A will be: (a) 6.37 A (b) 7.071 A (c) 10 A (d) 10.67 A.
24. The capacitive reactance of a 100 µF capacitor connected to the mains supply will be: (a) 0.314 Ω (b) 31.83 Ω (c) 5000 Ω (d) 31 kΩ.

25 The inductive reactance of a 0.10 H inductor connected to the mains supply will be: (a) 0.314 Ω (b) 31.42 Ω (c) 31.83 Ω (d) 3142 Ω.

26 Two a.c. voltages V_1 and V_2 have values of 20 and 30 V respectively. If V_1 leads V_2 by 45° the resultant voltage will be: (a) 16 V at 24° (b) 45 V at 90° (c) 46 V at 18° (d) 50 V at 45°.

27 The transformation ratio of a step-down transformer is 20:1. If the primary voltage is 240 V the secondary voltage will be: (a) 2.4 V (b) 12 V (c) 20 V (d) 24 V.

28 A sinusoidal alternating voltage has a maximum value of 340 V. Draw to scale one full cycle of this waveform and find (a) the instantaneous value of the voltage after 60°, (b) the rms value of the voltage using the mid-ordinate rule and (c) the average value of the voltage using the mid-ordinate rule.

29 Describe the structure of a material which is classified as: (a) a good conductor and (b) a good insulator.

30 Describe with sketches the meaning of the terms *frequency* and *period* as applied to an a.c. waveform.

31 Describe with the aid of phasor diagram sketches, the meaning of (a) a bad power factor, (b) a good power factor and (c) explain how a bad power factor may be improved.

32 Describe the construction of a small transformer designed to reduce losses when operating (a) at mains frequencies only and (b) at radio frequencies.

33 Sketch the voltage and current phasor diagrams for a series circuit containing:
 (a) resistance and inductance
 (b) resistance and capacitance
 (c) resistance, inductance and capacitance.

34 Sketch the phasor and waveform diagrams for a series circuit at resonance.

35 State one advantage and one disadvantage of a series resonant circuit.

36 Calculate the resonant frequency when an inductor of 1 mH is connected in series with a 1 µF capacitor.

37 Use a neat sketch to show how an ammeter and voltmeter are connected to a simple series circuit to measure total current and total voltage.

38 Calculate the turns ratio of a matching transformer to ensure maximum power transfer from a 5k Ω amplifier to an 8 Ω speaker.

9

ELECTRONIC SYSTEMS

The word *system* has only recently become popular and assumed importance as a result of the widespread use of the term. Our world is a part of the *solar system*, we live in a *social system*, doctors study the *nervous system* and scientists study the *ecological system*. In general, the theory of systems looks at the things which are common or related in an effort to make the overall behaviour more understandable. This moves us towards a definition of the system. 'A system is a collection of parts which are joined or connected together in some particular way.'

A systems approach is a procedure or strategy which is used as a way of finding solutions to a complex problem. This involves using a system diagram or block diagram linked by interconnecting lines and arrows.

A *system* or *block diagram* is made up of a number of boxes, each box representing a part of the system which can also be called a sub-system. The purpose of the block diagram is to show how the various parts or sub-systems relate to, or interconnect with, each other. These boxes are often called *black boxes* because initially we don't want to see what is inside the box. We only want to understand the overall behaviour and the relationship of one box with another.

The interconnections between the various parts of the system, or between the black boxes, are represented by lines and arrows, the arrows indicating the direction of flow.

This approach to problem solving allows us to concentrate our efforts on understanding the system without getting confused by the circuit complexities of each black box.

A cassette tape recorder can be used to store sound on a magnetic tape and, some time later, to play back those sounds through a speaker. A tape recorder is a sophisticated piece of electronic equipment made up of complex circuits, but if we look at it as a system which is made up of black boxes, we can begin to understand how it works without necessarily being able to understand the individual parts of the circuit. Fig. 9.1 shows a system diagram for a cassette tape recorder.

The first black box represents the microphone which picks up the sound waves and converts them into an electrical signal. The amplifier, box two, modifies those signals and stores them on the magnetic tape, box three. On playback the magnetic tape signal must be amplified, box four, before being played back through the speakers. Therefore, the tape recorder has five sub-systems or black boxes. Each box has a different function and in reality will be made up of complicated electro-mechanical components and circuits. Even though we don't yet know how each block will be constructed, we have made a start in our understanding of how the tape recorder works. A great deal can be done in electronics if you know what a particular black box should do, even though you don't know how it does it.

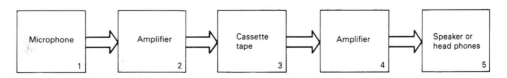

Figure 9.1 A system diagram for a tape recorder.

In electronics, engineering, instrumentation and control, we are usually using an electronic system to control something for a purpose, for example, to control water temperature or motor speed or fluid levels or the output of an audio system. Whatever their level of sophistication, all control systems have certain common basic features. The simplest form of control is open loop control.

Open loop control

Open loop control involves designing a system to do a particular job as carefully as possible and then leaving it to work on its own. With open loop control you are not in a position to make corrections once the system is set up and working. Consider the following example. A golfer stands suitably poised, holding the club and ready to take a shot. He addresses the ball and, after swinging back the club with great care, strikes the ball. When the ball has left the club face the golfer has lost all control over the flight of the ball and further corrections cannot be made. If the shot is not going where it was planned it should go, then all that can be done is to take another ball and try again. This is, therefore, an example of open loop control.

Consider the d.c. power supply shown in Fig. 9.2. The transformer reduces the a.c. mains voltage to a lower a.c. voltage. The rectifier converts this a.c. voltage into a d.c. voltage by connecting four diodes in a bridge circuit. The output from a bridge circuit is lumpy d.c. and therefore the filter is required to smooth the output. This will probably be an electrolytic capacitor connected across the output. The d.c. power supply will have been carefully designed and the components carefully selected to give a specified output, but if any of the variables change, there is no facility to make corrections and, therefore, this too is an example of open loop control.

Closed loop control

For closed loop control there must be a way of feeding back information so that adjustments can be made to correct errors in the system. Driving a motor car is a good example of closed loop control. The driver constantly makes corrections to speed and direction in response to observations of the changing road conditions.

MOTOR SPEED CONTROL

The motor speed controller shown in Fig. 9.3 is also an example of closed loop control. The error detector has two inputs and one output. The desired speed is set by the speed control which supplies a reference signal to the error detector. The output from the error detector supplies a signal to the power amplifier, which provides the necessary power to drive the motor. The drive shaft of the motor, in addition to driving the mechanical load, also turns a tacho

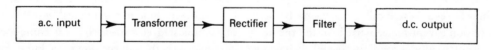

Figure 9.2 d.c. power supply.

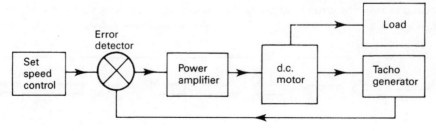

Figure 9.3 Motor speed control.

generator. This is a small generator coupled to the drive shaft, which generates a voltage which is proportional to the shaft speed. Increasing the shaft speed increases the generated voltage, reducing the speed reduces the generated voltage. This voltage is then fed back to the error detector, which compares the feedback signal with the reference signal. The difference between the reference signal and the feedback signal, the error, is then used to adjust the output to the power amplifier, so that the chosen speed can be maintained.

This system is, therefore, made up of five black boxes or sub-systems. The set speed control, which provides the reference signal, might be a voltage set by the wiper of a variable resistor or potentiometer. The error detector could be an operational amplifier because op amps can compare the voltage at their two inputs. The power amplifier will probably be an electronic circuit built up from discrete components as shown in Fig. 9.4 and considered in more detail in Chapter 5. Low power systems can be made into integrated circuits and cooled by mounting on heat sinks, but high power systems tend to be made with discrete components because they generate a lot of heat and need to be bulky to provide the necessary cooling of the components. The power amplifier drives the motor which, in turn, drives the tacho generator. This generates the feedback voltage, which is connected to the input of the error detector, forming the closed loop in this system.

The feedback signal of a closed loop system can be either positive or negative.

Figure 9.4 Using heat sink and discrete components to cool a power amplifier.

Negative feedback

Negative feedback is the name given to the situation in which the feedback signal is subtracted from the reference signal with the intention of reducing the error or variations in the system. Negative feedback is the type normally used in control systems because it leads to stability. The motor speed control system described above and shown in Fig. 9.3 will incorporate negative feedback because the system is designed to reduce variations in the speed of the motor. Electronic amplifier circuits normally incorporate negative feedback because the voltage gain of the amplifier can be accurately determined by the value of the feedback components. The frequency response of the amplifier is also improved, that is a wide range of frequencies are amplified by the same amount and the output of the amplifier, although increased or amplified, is a true copy of the input with little distortion.

Positive feedback

Positive feedback is the name given to the situation in which the feedback signal is added to the reference signal. This usually leads to instability and can be used in electronics to generate oscillations for use as the output of signal generators.

INSTABILITY

Instability can occur in any closed loop system because the output of each black box in the system forms the input to the next box. When a controlling action is required, the necessary changes can take some time to work their way through each block in the system from input to output. Let us suppose a reference sine wave was applied to one of the blocks in a system as shown in Fig. 9.5. With negative feedback applied, the feedback signal will be negative with respect to the reference signal at any instant in time. The feedback signal will, therefore, be subtracted from the reference signal, resulting in stability of the output. If there is a time lag in the system, resulting in a phase lag of say 180°, the feedback signal will be in-phase with the reference signal, resulting in positive feedback and instability of the

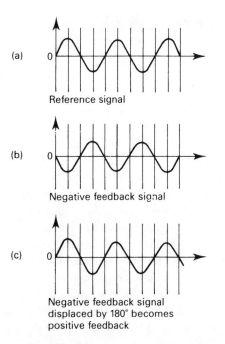

Figure 9.5 Effects of feedback.

system. An example of this effect is the howling of a loudspeaker when it is brought close to a microphone. The sound from the loudspeaker is picked up by the microphone but a time delay occurs as the sound waves travel through the air between the speaker and the microphone. This signal lag causes some frequencies to be amplified, resulting in the howl. Greater separation of speaker and microphone resolves this particular problem.

Transducers

A transducer is a device which converts a particular physical output into another type of physical output. The microphone and loudspeaker are both transducers, the microphone converts sound waves into electrical signals and the loudspeaker converts electrical signals into sound waves. Other transducers are considered in Chapter 12. The output from the transducer is called the signal and the *signal conditioner* converts this output into a form which is compatible with the electrical circuits. The tachogenerator of Fig. 9.3 converts speed of rotation into an electrical signal, for feedback to the operational amplifier acting as the error detector. In control systems the transducer often acts as a *sensor*, providing signals which allow a circuit to be controlled in a predetermined way. Other examples of sensors are solenoids, magnetic valves, pressure switches, microswitches, thermocouples and thermostats.

WASHING MACHINE CONTROL

A washing machine control system is made up of a number of sub-systems, many of which are sensors providing inputs to a management sub-system or controller. A typical system is shown in Fig. 9.6. The water level sensor would typically be a small ball valve-operated microswitch which would switch off the water supply when a pre-determined level of water in the drum was reached. The water temperature sensor would be a thermostat switching off the electrical supply to the heater elements when a suitable washing temperature has been reached. The drum motor rotates the drum which contains the clothes and water, usually turning in one direction for a short time and then in the reverse direction to prevent the clothes becoming tangled. The pump is a water pump, removing waste water from the drum after each wash and rinse operation. The controller/timer is often a spring action rotary timer which incorporates a number of switch contacts or a semiconductor logic control device. This makes it possible to sequence the various operations such as fill the drum to the required level, raise the temperature of the water before rotating the drum and clothes. After a pre-determined time has elapsed, the pump removes the dirty water from the drum, the pump motor stops and clean water flows into the drum to the pre-determined level, before the drum is once more rotated for the rinsing cycle. Upon completion of the rinsing cycle the water is pumped out and the clothes are then ready for drying. All these operations need to

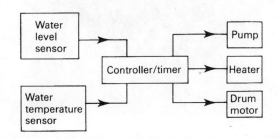

Figure 9.6 A washing machine control system.

be timed and many interconnections must be made. The controller/timer is the sub-system which manages these functions. If it becomes faulty the whole system will fail to operate. If the pump motor fails to pump, the drum will remain full of water. If the heater element fails, the clothes will be washed in water which is below the desired temperature. If the drum fails to rotate, the drum motor or drive belt may be broken and the clothes will simply soak in the water without being tossed about, which provides the best cleaning action. If the water temperature thermostat becomes open circuit, the water heater will not be switched on and the clothes will be washed in cold water. If the thermostat remains closed, the heater will remain switched on and the clothes will be washed in very hot water which may damage wool or delicate fabrics. If the water level sensor becomes faulty, the water will either not begin to fill or fill continuously until water overflows. The controller/timer permits some degree of sophisticated control by the interconnection of the various circuits and sensors. For example, the water heater will only operate when the correct level of water has been reached, and even then, only if the water is below the desired temperature.

Security systems

In recent years we have seen an increase in the installation of security systems to domestic and commercial premises for the protection of goods and property against the criminally inclined. The crime prevention authorities now all agree that the chance of a successful burglary is greatly reduced by fitting an effective alarm system. Fig. 9.7 shows the system diagram of a typical security system. All burglar alarms consist of three sub-systems; a warning device, detection devices and a control panel. The detection devices are the sensors of the system, which respond to the presence of an intruder or the opening of a window. The sensors trigger an audible warning device, usually an alarm bell and flashing light. The system is switched on and off at a control panel with a key or coded buttons. The control and timing unit is the master control panel for the whole security system. All the system components are joined together with small, low-voltage cable which is easily concealed.

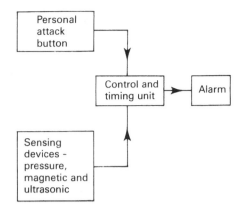

Figure 9.7 A security control system.

The sensing devices, which detect the presence of a burglar, might be proximity switches on doors and windows or movement detectors such as passive infrared or ultrasonic detectors. A personal attack button is a press switch which is used to activate the alarm system manually in situations where the authorised occupants of the building become aware that an intruder is present. A personal attack button is shown in Fig. 9.8.

Security system false alarms are most annoying to neighbours and the police. Water ingress can cause proximity switches to fail, movement sensors can be activated by domestic pets. However, by carefully selecting the equipment and adjusting the detector's sensitivity, false alarms can be minimised so that the system gives many years of trouble-free service and peace of mind. Security systems are further considered in Chapter 11.

Figure 9.8 A personal attack button.

Space heating control

A space heating system is today as much a part of the total installation in a domestic or commercial building as the electrical wiring or the plumbing and sanitary fittings. With the acceptance of central heating has grown the awareness of the convenience and cost effectiveness of controlling the system. In the 1950s a basic time clock was available to give simple on/off control but the technical advances made in the engineering of heating systems, particularly the introduction of flow valves and wholly pumped systems, has led to complete energy management systems in the 1990s. Programming and sensing units coupled to flow valves can give maximum economy and comfort for a selected temperature range at any time of the day or night. A typical gas-fired or oil-fired space heating system is shown in Fig. 9.9.

The boiler provides the heat energy which is circulated around the system as a result of burning fuel, either oil or gas in most domestic systems. The fuel burner always incorporates a flame failure device such as a thermocouple, so that the fuel flow is switched off if the flame should fail for any reason.

The fuel control valve is an on/off valve controlling the supply of fuel to the boiler burner which is operated electromagnetically. When the programmer and hot water or radiators call for heat, the magnetic valve is energised, opening the valve and allowing fuel to flow to the burner. The fuel control valve is designed to fail safe and therefore, a faulty valve would result in the boiler switching off. The programmer, room and water sensors would continue to call for heat but cold water only would be circulated because the boiler, the source of heat energy for the system, would have shut down as a result of the faulty valve.

Heat is transferred from the boiler, to the water heating cylinder or space heating radiators, by water contained in small bore pipes which are insulated to prevent heat losses. The water is transferred around the system by means of a water pump, usually a single phase induction motor which is energised when the programmer and hot water or radiators call for heat. If the pump fails, heat cannot be transferred to the hot water or radiators, but the boiler will ignite if the boiler thermostat is turned up when the programmer is switched on.

The controller can incorporate relays and connector blocks for the boiler, pump and valve switching; it is the management system responding to the input from the various sensors and activating the boiler fuel flow and pump. A three-way valve directs hot water from the boiler to either the hot water cylinder, the space heating radiators or both, dependent upon the settings of the programmer, room and cylinder thermostats. If the controller fails, the link between the inputs and outputs will be broken and the system will be inoperative.

The room thermostat is the room temperature sensor. If the room temperature is below the thermostat setting, the circuit is closed, activating the flow valve and heat is transferred, by pumping hot water from the boiler to the space heating radiators. The room temperature rises until the thermostat setting is reached, the switch then opens, closing the flow valve and switching off the pump.

The cylinder thermostat is the water temperature sensor, usually clamped to the hot water storage cylinder. If the water temperature is below the thermostat setting, the circuit is closed, activating the flow valve, and heat is transferred by the pump from the boiler to the water contained in the cylinder. The domestic hot water temperature is raised until the thermostat setting is reached, the switch then opens, closing the flow valve and switching off the pump.

The programmer incorporates a time clock to determine when the system is switched on or off and a selector switch which allows the user to predetermine what is required, space heating, hot water or both. With the programmer in the off mode, the whole system is shut down. With the programmer switched on, the system is designed to operate automatically, maintaining the water and room temperature at a predetermined setting through the flow valve, pump and temperature sensors.

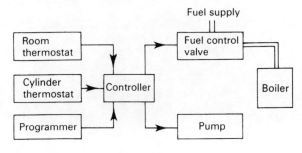

Figure 9.9 Space heating control system.

If the programmer fails, the most likely cause will be a faulty time clock, which will result in the whole system being either switched off or on continuously.

AM transmitter

The modulation, transmission and reception of an intelligence carrying signal is a complex technical operation but, if we look at it as a system, the technology becomes understandable. Figure 9.10 shows a system or block diagram for an AM (amplitude modulation) transmitter.

The RF oscillator generates a high 'radio frequency' signal carrier. The AF amplifier amplifies the small 'audio frequency' signals from the microphone. These two signals are then combined in the modulator (see chapter 5 under Signal modulation) before being amplified in the radio frequency (RF) power amplifier which provides the high output signal required by the aerial system.

The radio frequency energy is radiated into space from the aerial as an electromagnetic wave. At a distant receiving point, the radiated wave generates a voltage in the aerial system. However, the aerial will pick up many signals and the receiver equipment must, therefore, *select* the desired signal. Figure 9.11 shows a system or block diagram for an AM receiver.

AM receiver

The RF filter is a variable tuner which selects the desired signal from all the signals picked up by the aerial.

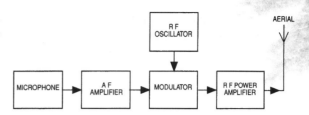

Figure 9.10 A system diagram for an AM transmitter.

The oscillator is also varied by the variable tuner controls but pre-set capacitors called 'trimmers' and 'padders' maintain the oscillator frequency at 465 kHz above the incoming signal frequency.

The oscillator output and the RF filter output are fed to the mixer which produces an output frequency equal to the difference between them. This is called the intermediate frequency or IF. The IF modulation is exactly the same as the selected RF signal and is amplified in the IF amplifier before being passed to the AM demodulator.

The AM demodulator separates the audio frequency intelligence from the IF signal and the resultant d.c. voltage is fed back to the IF amplifier as negative feedback to control the gain (or amplification) of the IF amplifier. This balances out the variations in signal strength which might be brought about by different transmitters and is called the automatic gain control or AGC.

The audio frequency signal is then amplified in a series of amplifiers passing first to the AF voltage amplifier and then to the AF power amplifier where the signal output is boosted sufficiently to drive the loudspeaker.

The IF amplifier can be designed to produce whatever bandwidth is required (see chapter 5 under

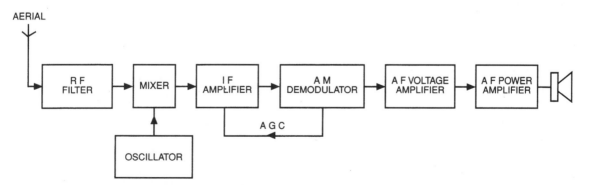

Figure 9.11 A system diagram for an AM receiver.

Band... of an amplifier'). The bandwidth will be ... small if the receiver is used for voice communica... only, wider for AM music reception, wider still ... gh quality FM reception and very wide for TV use.

Amplitude modulation uses comparatively simple modulation circuits and demodulation is easy, but much of the transmitter power is wasted and the peaks of the signal become distorted by interference with the result that some signal information is lost. However, it is also possible to modulate the *frequency* of the carrier wave so that it carries the required information. Frequency modulation or FM has a constant carrier amplitude which results in a greater transmitter range and more freedom from interference.

FM transmitter

Figure 9.12 shows a system or block diagram of a FM (frequency modulation) transmitter. The AF amplifier amplifies the small 'audio frequency' signals from the microphone. This signal is then used to modulate the *frequency* of a radio frequency carrier wave which is generated and modulated in the *frequency modulated oscillator*. The modulated radio frequency signal is then amplified in the RF power amplifier which provides the power required to radiate the signal from the aerial system.

FM receiver

Figure 9.13 shows the system diagram of an FM receiver. The FM signal is picked up by the aerial system of the FM receiver whose working operation is much the same as the AM receiver, both being based on the supersonic heterodyne ('superhet') principle. The RF amplifier is a tuned circuit which selects the desired signal from the aerial system and then amplifies it above the amplitude of any unwanted interference signals. It also prevents the oscillator frequency leaking back to the aerial system which would cause interference to other receivers.

The variable frequency of the oscillator output is controlled by the tuning of the RF amplifier being 'ganged' to it and, therefore, they are controlled together. The frequency of the oscillator is maintained at 10.7 MHz above the incoming signal frequency. This is because FM has a higher frequency range and greater bandwidth than an AM signal.

The oscillator output and the RF amplifier output are fed to the mixer which produces an output frequency equal to the difference between them and called the intermediate frequency or IF. The IF modulation is exactly the same as the selected RF signal and is amplified in the IF amplifier before being passed to the FM demodulator.

The FM demodulator separates the audio frequency intelligence contained in the frequency *changes* of the selected incoming signal. The resultant d.c. voltage is fed back to the oscillator in an FM system because correct tuning on the FM receiver depends upon the correct tuning of the oscillator frequency. This negative feedback balances out the variations in signal strength brought about by different transmitters and is called automatic frequency control or AFC.

The audio frequency intelligence is then amplified in a series of amplifiers passing first to the AF voltage amplifier and then to the AF power amplifier where the audio signal output is boosted sufficiently to drive the loudspeaker.

An FM receiver has the ability to 'lock on' to the slightly stronger of two signals of the same frequency and to reproduce only the modulation of the stronger signal. This efficient 'selectivity', freedom from interference and broad bandwidth make FM the preferred transmission for mobile radio and stereo broadcasting.

Cathode ray oscilloscope or CRO

The principle of operation and use of a CRO to measure voltage and frequency are described in Chapter 6. In this chapter we want to look at the separate parts of the CRO and its operation as a whole system. Figure 9.14 shows a system diagram of the CRO and Fig. 6.6 shows more details of the cathode ray tube.

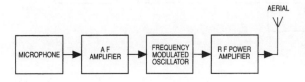

Figure 9.12   A system diagram for an FM transmitter.

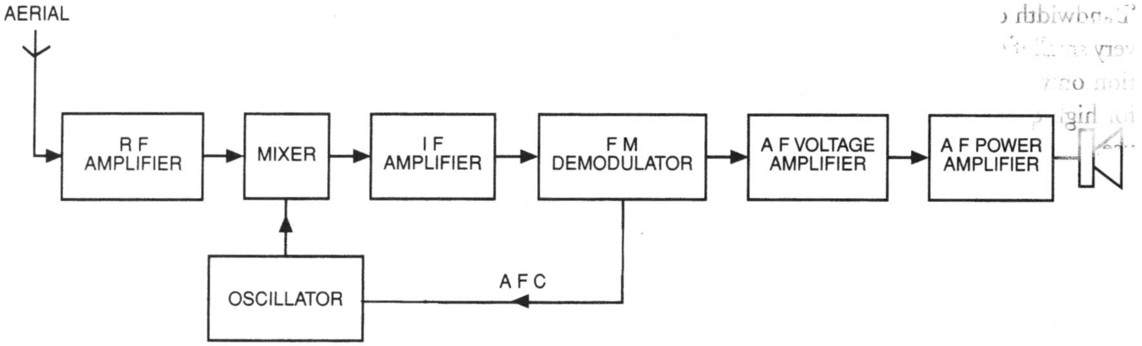

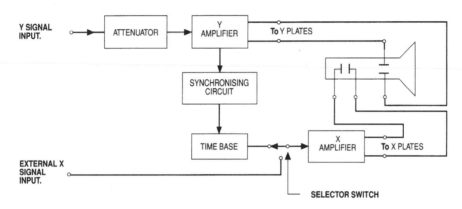

Figure 9.13 A system diagram for an FM receiver.

Figure 9.14 A system diagram for a CRO.

The signal to be displayed on the fluorescent screen of the CRO is applied to the Y-input connection. It is then passed to the attenuator where a large signal is reduced, and an amplifier, where a small signal is amplified. The amplifier and attenuator are matched and controlled by a switch calibrated in volts/cm. The input signal then passes to the Y-plates of the cathode ray tube where it is responsible for deflecting the electron beam in the Y-axis on the screen, that is, vertically.

A small part of the input signal is fed from the Y-amplifier to the synchronising circuit. The synchronising circuit is used to trigger the timebase signal at *exactly* the same time in each cycle so that a standing waveform can be displayed on the CRO screen. Without the synchronising circuit, the displayed waveform will appear to move across the screen to either the left or the right.

The internal timebase is a sawtooth signal generator which is fed to the X-amplifier and then to the X-plates where it causes the electron beam to deflect horizontally across the CRO screen. The sawtooth waveform generated by the timebase is shown in Fig. 5.37. During the rise time, the electron beam sweeps across the screen from the left to the right-hand side. During the short discharge time, the electron beam flies back to the left-hand side to begin its sweep once again.

The selector switch can be used to switch off the internal timebase so that an external signal can be connected to the X-plates. Lissajous figures are produced in this way and can be used to compare the frequency of the signal applied to the X-plates with that applied to the Y-plates.

Tape recorder

The modern tape recorder uses a fine plastic tape impregnated with powdered magnetic material which becomes re-arranged according to the strength and frequency of the signal being recorded. The magnetic tape retains the magnetic field pattern when the recording signal is removed and this can be used to set up a signal on playback which is identical to the signal which laid down the original magnetic pattern. This is the principle which can be applied to an audio tape recorder and Fig. 9.15 shows the essential building blocks of such a system.

The audio input is amplified by the 'record amplifier' so that a strong magnetic field can be developed in the 'recording head'.

Since the record and playback heads have essentially the same characteristics, most modern tape recorders combine them into a single unit. The head consists of a coil of wire wrapped around a magnetic core with a narrow gap where the tape moves across its surface. In the record position the head carries the record signal creating a 'unique' magnetic field at the air gap which is transferred to the tape as it passes. The tape moves on and the magnetic field is stored on the tape.

In the playback position the tape is passed across the gap in the magnetic head. The magnetic flux on the tape induces a corresponding flux in the magnetic core. This creates voltage signals in the coil wound around the core which are identical to the recording signals.

The erase head is constructed in the same way as the record/playback head but it is fed with an alternating current at about 20 kHz from the bias oscillator which produces a rapidly changing magnetic field. This shakes up the powdered magnetic material on the tape as it passes the head, erasing the magnetic field patterns made previously. In the record position, the tape passes the erase head immediately, before the record head so that the tape is 'clean' and ready to accept the new recording.

The bias oscillator also provides a bias signal to the audio input of the recording head. This overcomes the distortion which otherwise would occur because of the transfer characteristics of the magnetic tape.

The two switches are 'ganged' so that they operate together. In the record position the audio input is first amplified and then fed to the record head. At the same time the erase head and record head are connected to the bias oscillator.

In the playback position the erase head and bias oscillator are disconnected and the playback head is connected to the playback and AF amplifiers so that the recorded signal can be increased to the level required to drive the loudspeaker.

Digital clock

To measure time it is necessary to harness an event which occurs regularly and repeatedly. A grandfather clock uses the repeated event of a pendulum swinging

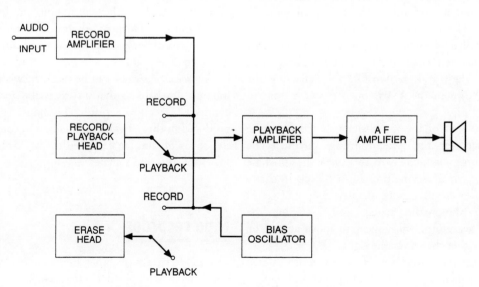

Figure 9.15 A system diagram for a tape recorder.

to and fro to drive the hour and minute hands around a clock face. The hours, minutes and seconds are divided up by the cogs of a gear train.

Most modern clocks and watches use a quartz crystal which can be made to vibrate at a regular interval if a voltage is connected across it (see Fig. 12.14 for the action of a quartz crystal).

The crystal vibrates very quickly and so a series of dividers are required before the oscillations can measure hours, minutes and seconds.

Figure 9.16 shows a system diagram for a digital clock. A digital clock is a clock which expresses time as a series of digits or figures. In electronics we usually use a seven segment display such as that shown by Fig. 4.11. The grandfather clock discussed earlier gives an analogue display because the fingers *point* to the value of time displayed around the clock face. The crystal oscillator is a quartz crystal combined with an electronic oscillator which produces a series of regular pulses. The frequency of the pulses is very high and, therefore, the output is fed through a series of dividers so that the number of pulses can be divided down.

The dividers are logic gates whose output will only switch on after a series of pulses have been received at the input. They are, therefore, acting as counters and are arranged to give one pulse per second, one pulse per minute and one pulse per hour.

The second, minute and hour signals are fed to a decoder and then on to a seven segment display. The decoder converts the logic inputs into the voltage required to make up the number to be displayed.

The dividers and decoders are generally packaged as integrated circuits (ICs) and combined divider–decoder and divider–decoder–display IC chips are available.

Computing systems

Modern microcomputers such as the personal computers found today in most homes, schools and offices perform logical operations on data in the form of digits (0 and 1) which have been converted into a series of electrical pulses. They are built from a large number of logic gates which can be programmed to provide different pathways, allowing the computer to perform in various ways. The 'power' of a computer depends upon the speed of operation and the amount of data it can handle simultaneously.

All physical components are called *hardware*. Examples of hardware are the keyboard, the printer, the visual display unit (VDU), the disc drive and the joystick. The programs that run on the system are termed *software*. These programs can be altered, amended or linked and are, therefore, 'soft' in terms of their long term operation.

Other components within the system such as floppy discs or CD ROMs are a little more difficult to categorise. These are physical objects which would imply hardware but they also contain programs which implies software. These devices are, therefore, sometimes referred to as *firmware*.

The computing system is built up of four sub-systems as shown in Fig. 9.17. The basic functions are input, central processing and output, with the various processing stages being controlled by the computer's memory.

The input device reads the input data and converts it into a form of 0s and 1s that the computer will understand. Typical input devices include keyboards or voice recognition firmware, magnetic tape or compact discs (CDs) or analogue transducers such as

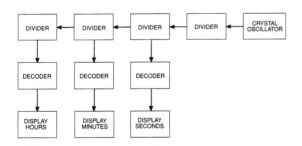

Figure 9.16 A system diagram for a digital clock.

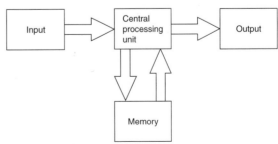

Figure 9.17 A system diagram for a computer.

level, speed or heat detectors as described in Chapter 12 of this book. Analogue signals are unable to generate the correct sequence of 0s and 1s which the computer recognises and, therefore, an *interface* is required between the input and central processing unit to convert the analogue signal into a digital signal.

The central processing unit (CPU) is the computer's brain. It accepts signals from the input unit which then interact with the computer's programs to provide the required output. Personal computers usually advertise the type of CPU being used by that particular PC somewhere in the specification. Pentium III processors are an example of the IBM style of computer being used extensively these days in the home, school and businesses. This quality product gives an indication of the power and potential speed of the particular PC.

The CPU contains three important sections: a control unit, an arithmetic and logic unit and an accumulator. The control unit contains a crystal controlled clock to generate timing pulses and a program counter to point the way to the address of the next instruction. The arithmetic and logic unit performs arithmetic calculations in either binary arithmetic or binary logic form. The accumulator contains the data actually being processed before it is passed to the memory or the output.

Output devices enable the computer to communicate with or exert some sort of control over industry or commerce. They include devices such as VDUs or monitors, printers and computer controlled robots for motor car assembly. Robots are usually controlled by stepper motors whose principle of operation is discussed in Chapter 12 of this book.

A computer also needs a memory to store the program of instructions which it needs to operate in addition to the input data that is waiting to be processed. This requires two types of memory, ROM and RAM. Read only memory (ROM) retains its information even when the power is switched off and it is used to carry the operating instructions of the computer. Random access memory (RAM) is capable of being written to and read from in an almost random manner. RAM is required to store the data for processing because it needs to change to allow instructions to be written in, altered and read out. Because it has a retentive memory, ROM is described as being *non-volatile* while RAM, which loses the data when the power is switched off, is said to be *volatile*.

When the computer is switched on, data is read from the hard drive on to the RAM, where it is modified, altered or updated and then 'saved' on the hard drive before the power is switched off. The hard drive consists of many magnetic discs rotating on the same shaft. Very small electromagnetic heads can be placed on the discs to read or write data. A hard drive will typically have a capacity of between 2.5 Gb and 10 Gb these days.

Personal computers also use floppy discs and CD ROM drives to store data. A 3.5 inch floppy disc has a capacity of 1.4 Mb. CD ROM drives have a capacity of 650 Mb and modern drives permit both reading and writing facilities.

DVD or digital versatile disc is a further development of the CD (compact disc). DVD has a much greater capacity to store information, currently 17 Gb, and can be used for any type of digital data such as motion pictures, 17 hours of music per disc in addition to computer data storage.

If you were to spend about £1000 today (2000) on a computer package from one of the high street household name shops, your PC would probably have 64 Mb of RAM, a 6.4 Gb hard disc and incorporate floppy disc, CD ROM and DVD drives. The new millennium will see household PCs increasingly use DVD in read and write form and have 10 Gb hard drives to increase the computer's internal memory and speed of operation.

Drawings and diagrams

In electronics, as in many other engineering disciplines, we communicate many of our ideas and information with diagrams. Various types can be distinguished: block diagrams, wiring diagrams, circuit diagrams and supplementary diagrams. The one to use in a particular application is the one which most clearly communicates the desired information.

A BLOCK DIAGRAM

A block diagram is a relatively simple diagram in which an installation or piece of equipment is represented by block outlines. The purpose of a block diagram is to show clearly the operation or function of a subject or system and, therefore, it does not normally show the physical layout of any components or the individual circuit connections. Figures 9.6, 9.7 and 9.9 to 9.16 are examples of block diagrams.

A WIRING DIAGRAM

A writing diagram or connection diagram shows the detailed connections between components or items of equipment and in some cases, the routeing of these connections. The purpose of a wiring diagram is to help someone with the actual wiring of the circuit. Figure 9.18 is an example of a wiring diagram.

A CIRCUIT DIAGRAM

A circuit diagram shows most clearly how a circuit works. All the essential parts and connections are depicted by their graphical symbols. The purpose of a circuit diagram is to help in the understanding of the circuit and it should be laid out as clearly as possible without regard to the physical layout of the actual components or parts. Figure 9.19 is an example of a circuit diagram, as are most of the diagrams in Chapter 5.

A SUPPLEMENTARY DIAGRAM

A supplementary diagram does not fall into any one of the above categories. Its purpose is to convey additional information in a form which is usually a mixture of the other three categories. Figure 9.20 shows a typical supplementary diagram, the cabling arrangements for a space heating system.

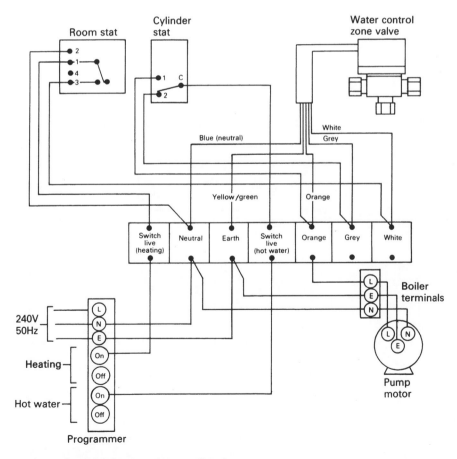

Figure 9.18 Wiring diagram for space heating control (Honeywell 'Y' plan).

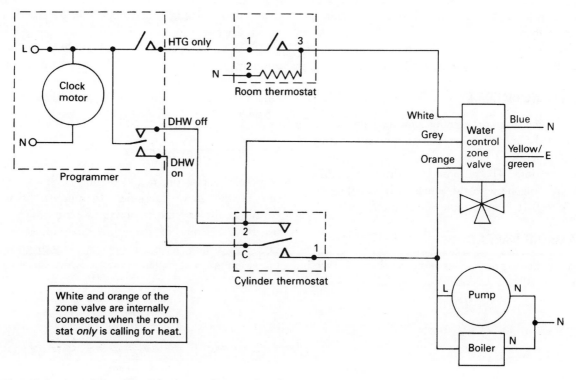

Figure 9.19 Circuit diagram for space heating control (Honeywell 'Y' Plan).

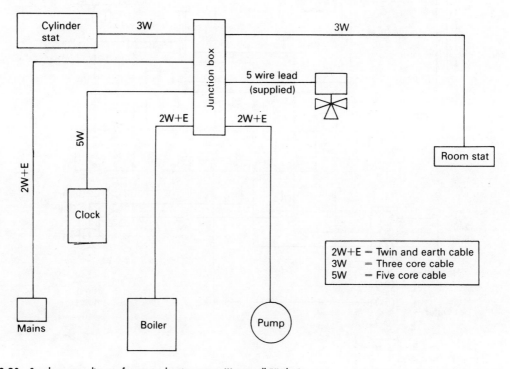

Figure 9.20 Supplementary diagram for a space heating system (Honeywell 'Y' plan).

COMMUNICATION SYSTEMS

Simple communication systems

The purpose of any communication system is to convey information between two physically remote points. Any communication system must include a transmitter for sending the information, a transmission circuit or medium and a receiver. To transmit signals comprehensible to the human ear the system must produce an audible note. The simplest system is a morse code or buzzer which produces a sound whenever a key activates the circuit. This sound may be of one frequency only but for the transmission of speech a complex mixture of frequencies is required and, therefore, the communication system must be capable of accommodating a wide range of frequencies. The range of frequencies required by the communication system is called the bandwidth: the simpler the system, the smaller the bandwidth.

The bandwidth of speech is from 30 Hz to 5 kHz. Music requires an even greater bandwidth from 20 Hz to 20 kHz. The telephone system is designed to transmit on a limited bandwidth from 300 Hz to 3.4 kHz. This is suitable for acceptable voice reproduction but, because the upper and lower frequencies are cut off, the quality of reproduction is not perfect.

Simple telephone circuit

A telephone handset contains a receiver, the ear piece and a transmitter for speaking into. The transmitter converts the pressure waves from the spoken word into analogue electrical signals, which are transmitted along conductors to the receiver of another handset which converts the electrical signals into sound waves. This provides the basis of a simple two-way communication system and a suitable circuit is shown in Fig. 10.1.

The modern telephone system

One of the limitations of the simple telephone circuit is that a separate cable is required for each telephone conversation. If there are more people in London wanting to call someone in Edinburgh than there are lines between London and Edinburgh, then someone will have to wait. This problem can be resolved by putting up more lines, as was the case before 1940 when main roads and railways were lined with telegraph poles carrying dozens of wires, or alternatively, more than one conversation can be sent down each line. This is the modern method which is called *multiplexing*. Each conversation is put

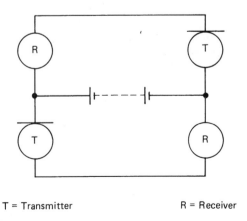

T = Transmitter   R = Receiver

Figure 10.1 A simple telephone circuit.

on to a carrier wave of a different frequency and all these different frequencies are sent down the line at the same time without interfering with each other. At the receiving end a filter is used to pick out a particular frequency and, therefore, a particular conversation. Another receiver can be tuned to a different frequency and, therefore, select a different conversation.

Optical fibres

For a long time telephone lines have used copper cables, but these will eventually be replaced by optical fibres which can carry more conversations with reduced line losses. Line loss is called attenuation.

Glass fibre cables are about 0.1 mm in diameter and trap the light signals inside the very pure optical-quality glass fibres, using total internal reflections. The laser light signals bounce their way along the cable at the speed of light, always being reflected when they strike the inside of the cable wall in the same way that light is reflected in the prisms of binoculars. The optical-quality glass fibres are so pure that a 2 km length of cable will absorb less light than a sheet of window glass.

Some attenuation (line loss) does occur and eventually the signal strength will decrease. To prevent this, amplifiers called repeaters or boosters are installed along the route, about every 30 km of line length, to restore the signal strength. Copper cables require repeater stations much closer together. The repeaters, while restoring the required signal strength, unfortunately also amplify the unwanted random noise in the system which sounds like a hiss or crackle in the receiver. In some situations, such as on long lines, the sound quality can deteriorate until the conversation is incomprehensible. This random, unwanted noise can be eliminated by using digital signals as was discussed in Chapter 8.

System X

This is the name given to a new British Telecom telephone system using digital electronic exchanges which was introduced in 1978 and will eventually replace all of the existing exchanges. All signals will be handled by an *integrated digital network* computer system. Speech signals will be transmitted in digital form. The analogue voltages generated by voice patterns will be converted to digital form by analogue-to-digital converters. All signals entering the exchange, voice, computer and fax machine data, telex, Datel and Prestel will be handled by the same computer-type circuits. The development of this system will mean that the exchange of all types of information over the telecommunications network will be faster and simpler.

Microwave telephone links

Microwaves are radio waves with a large capacity to carry information. They have a very small wavelength and can, therefore, be easily focused into a narrow beam by a dish aerial of about 2 to 3 metre diameter. Almost half the trunk calls in the UK are transmitted by terrestrial microwave links which cross the country in line-of-sight hops between dish aerials mounted on tall towers built on hilltops. The received signal is boosted or amplified before being transmitted on to the next hop. The British Telecom tower in London is the nerve centre of this system.

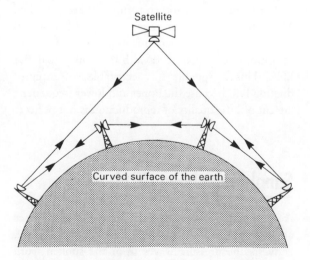

Figure 10.2 Microwave links between line of sight aerials and between earth stations linked by satellite.

Satellite telephone links

Satellites can be used to transmit signals between subscribers on different continents. Two thirds of all intercontinental telephone calls are now handled by the satellite network. A typical intercontinental call goes from the caller to a local exchange where it is routed to an international exchange and then on to an earth station. The earth station transmits the signal by microwaves up to the satellite where it is retransmitted to an earth station in another country. The call is then routed via international and local exchanges to make the link between the called number and caller.

Despite the expense involved in building the satellite and placing it in orbit, it can be cheaper than a network of transmitters and receivers which can 'see' each other over the curved surface of the earth.

Satellite communications are also discussed later in this chapter under the sub-heading Geo-stationary satellite communications.

Table 10.1 Telephone cable identification

code	base colour	stripe
G.W	green	white
B.W	blue	white
O.W	orange	white
W.O	white	orange
W.B	white	blue
W.G	white	green

Table 10.2 Telephone socket terminal identification

Terminals 1 and 6 are frequently unused and therefore 4 core cable may normally be installed.
Terminal 4 on the incoming exchange line is only used on a PBX line for earth recall.

Socket terminal	circuit
1	spare
2	speech circuit
3	bell circuit
4	earth recall
5	speech circuit
6	spare

Mobile telephones

These are also called cellular telephones because the country is divided into small areas or cells. Each cell has its own low-powered radio transmitter which is linked through regional exchanges to the existing telephone system. When a caller travels from one cell to another, a computer system switches the signal to the next cell so that the call can continue uninterrupted.

Telephone at home

The installation of telecommunication equipment could, for many years, only be undertaken by British Telecom engineers. However, this monopoly was relaxed by HM Government from January 1985, which created potentially new markets for the retailer to supply, and the electrical contractor to install, telecommunication equipment.

On new installations the electrical contractor or competent installer may install sockets and the associated wiring to the point of intended line entry, but the connection of the incoming line to the installed master socket must only be made by a BT engineer.

On existing installations, additional secondary sockets may be installed to provide an extended plug-in facility. Any number of secondary sockets may be connected in parallel but the number of telephones which may be connected at any one time is restricted.

Each telephone or extension bell is marked with a ringing equivalence number (REN) on the underside. Each exchange line has a maximum capacity of REN 4 and, therefore, the total REN of all the connected telephones must not exceed 4 if they are to work correctly. If REN 4 is exceeded the volume of one or all of the ringing devices will be reduced or they may not work at all.

An extension bell may be connected to the installation by connecting the two bell wires to terminal numbers 3 and 5 of a telephone socket. The extension bell must be of the high impedance type having a REN rating. All equipment connected to a BT exchange line must display the green circle of approval.

The multicore cable used for wiring extension socket outlets should be of a type intended for use

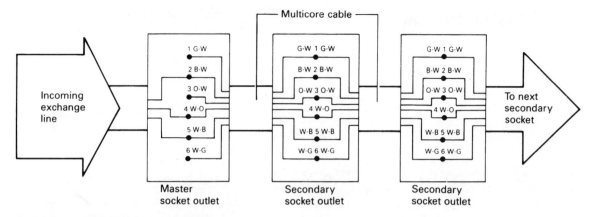

Figure 10.3 Telephone socket outlet connection diagram.

with telephone circuits, which will normally be between 0.4 mm and 0.68 mm in cross-section. A four-core cable should be run from the master socket outlets to each subsequent secondary socket outlet in the form of a radial circuit, and connected as shown in Fig. 10.3.

Radio transmission

When an electric current flows in a conductor a magnetic and electric field will also be established in the vicinity of the conductor. If an alternating current is flowing in a conductor, the magnetic and electric fields will also attempt to alternate. When the current reverses direction, the magnetic field must first collapse into the conductor and then build up in the opposite direction. However, it takes time for the magnetic field to collapse and then build up again, and at frequencies above 15 kHz not all the energy contained in the field has time to return to the conductor. The energy left outside the conductor cannot return to it and is radiated away as an electromagnetic wave travelling at the velocity of light, that is 3×10^8 m/s or 186 000 miles per second. The amount of energy radiated from a conductor, or aerial system, increases with an increase in frequency because more energy is unable to return to the conductor.

It was James Clerk Maxwell, a brilliant mathematician, who suggested in 1864 that electromagnetic waves must exist. In 1886 Heinrich Hertz became the first to generate these mysterious, invisible waves, but neither he or any other scientist of the day could suggest a practical use for them. It was in 1896 that Marconi demonstrated his wireless apparatus (without wires) and transmitted signals across the Atlantic ocean. Marconi was puzzled how radio waves could apparently travel around the curvature of the earth because Hertz had shown that electromagnetic waves travel in straight lines and, therefore, in theory, they should leave the earth's surface just beyond the horizon and vanish into space. This problem was resolved in 1902 by Arthur E. Kennelly and Oliver Heaviside who suggested that an ionisation layer existed above the earth which acted like a mirror reflecting the waves back down to earth again. This theory was proved experimentally in 1920 by Sir Edward Appleton and it is now accepted that long distance radio transmissions bounce off the reflecting ionosphere back to the surface of the earth.

Radiated energy reaches the receiving aerial system by one or more of three main methods of propagation, a surface wave, a space wave or a sky wave, as shown in Fig. 10.4.

The surface wave is guided around the earth by following a channel formed by the surface of the earth and the reflecting ionosphere.

Space waves travel in straight lines between the transmitter and receiver. If the receiver is not in 'line of sight' transmission by space wave alone is not possible.

Sky waves are radiated directly into the upper atmosphere where, depending upon the frequency and angle of incidence, they are reflected back to earth by one of the layers of charged particles collectively known as the ionosphere.

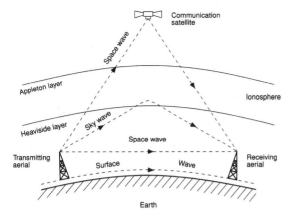

Figure 10.4 The three main methods of transmitting radio waves.

placed on high ground, but modern technology has also made it possible to place satellites in space and to use these to reflect microwaves back to earth. In this way microwave transmissions can cover a large area of the earth's surface as shown by Figs. 10.2 and 10.6. The microwave frequency range is used for radar, telecommunications, direct TV broadcasts and communication with weather and television satellites.

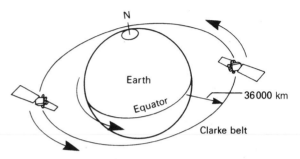

Figure 10.5 Satellites in geostationary orbit.

At low and medium frequency transmissions (30 kHz to 3000 kHz) the height of the transmitting aerial is small compared with the signal wavelength and transmission is mostly by ground waves. This frequency range is used for sound broadcasting and maritime communications.

High frequency transmissions (3 MHz to 30 MHz) are mostly propogated by sky waves bouncing off the layers of the ionosphere and the surface of the earth. World wide reception is possible by making several bounces, the distance travelled in one bounce, from transmitter to ionosphere and back to earth again is called the 'skip distance'. Distances of between 2000 km and 4000 km can be achieved in one skip depending upon the frequency used and the weather conditions. The high frequency range is used for short wave radio broadcasting, amateur and CB broadcasting, maritime and aircraft communications and international telephony.

Very high frequency (VHF – 30 MHz to 300 MHz) and ultra high frequency (UHF – 300 MHz to 3000 MHz) transmissions make use of the space wave for line of sight communications because at frequencies above 30 MHz the sky wave penetrates the ionosphere and disappears into space.

The VHF and UHF frequency band is used for commercial FM radio and television transmissions, the police, fire and ambulance emergency services, aircraft landing and telephone communications.

At super high frequencies (above 3000 MHz) the wavelength becomes extremely small, in the microwave region and transmission is only possible by line of sight space waves. A maximum range of about 150 km is possible by terrestial transmitting aerials

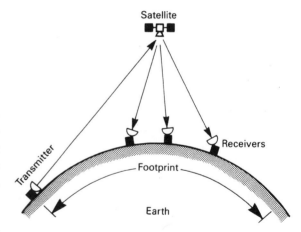

Figure 10.6 Using satellites to communicate around the curved surface of the earth.

Geostationary satellite communications

A satellite placed in a circular orbit around the earth at a distance of 36 000 km (22 300 miles) above the equator will appear stationary to an observer on earth if the orbital speed of the satellite is 'matched' to the earth's surface speed.

These are the conditions for an earth stationary (geostationary) orbit described in 1945 by Arthur C. Clarke, the British writer and scientist.

In this orbit the satellite will always remain over the same point on the surface of the earth. If the satellite is equipped with a radio transmitter it can *illuminate* a large area of the earth with radio waves which is called its *footprint*. Within this footprint earth stations can communicate with each other via the satellite at any time during the day or night.

Large earth station dish antennae of up to 30 m diameter are used for telecommunications and TV distribution. Medium-sized earth stations use 3 to 5 m dishes for the transfer of commercial data such as telex and facsimile transmissions. Small dishes, up to about 80 cm diameter, are used for the reception of TV signals broadcast by satellite directly to individual homes.

Satellite television

It is generally agreed that satellite TV will bring about one of the biggest advances in home entertainment since the BBC was formed in 1922. Commercial analysts calculate that 10 million homes in the UK already have their own satellite TV system and, therefore, the installation of satellite dishes represents a market with enormous growth potential for any competent installer.

A number of English language TV channels are transmitted from the Astra satellite, for example B Sky B, Sky News, Sky Movies, Sky Sport and UK Gold. Any of these TV channels can be received by pointing a receiving dish at the Astra satellite. The broadcasters using satellites are commercial profit-making companies and, therefore, the transmitted signal is scrambled. Individual subscribers must purchase a dish aerial, a receiver and a decoder into which the viewer must insert a credit card, paying a monthly subscription fee to unscramble the signal.

A fixed dish will only receive the signals from one satellite, that is, the one at which it is pointing. To receive the TV channels from other satellites would either require another fixed dish or one motorised dish which can be pointed at any satellite. A fixed dish pointed at the Astra satellite will give English speaking viewers access to a large number of TV channels in addition to the five channels already offered by the BBC and ITV companies.

Satellite dish installation

The installation of a satellite dish would at first glance seem to require a degree from NASA in mathematics, geometry and navigation! The installation instructions talk of azimuths, lines of longitude and latitude; orbits and inclinometers. The practical considerations are, in fact, much simpler. Installation is ultimately a matter of pointing the dish at the satellite and making sure it doesn't fall off the wall.

SITE SURVEY

The satellite is parked over the equator in a stationary orbit and, therefore, any dish antenna mounted in the UK must point south towards the satellite. The Astra satellite is to the east, or left of a reference line pointing true south, as shown by Fig. 10.8. The satellite dish must have a clear line of sight to the satellite position because microwaves will not pass through buildings, fences or trees. It must not be mounted above the line of the roof because this will probably break the local planning regulations.

DISH ASSEMBLY AND FIXING

Upon arrival at the site, the dish should be assembled. The feedhorn assembly usually is held at the focal length of the dish by a single arm or three rods. This distance is critical and the dish must be assembled exactly as instructed by the manufacturer. When the dish has been assembled, check for dish warp. Dish warp and incorrect focal length are common causes of poor reception. The easiest way to check for dish warp is to look across the dish from the front edge to the back. Each dish should be perfectly aligned. However, if the dish is warped, first check the fixing of the support rods, and if these are correctly attached to a metal or aluminium dish, the distortion may be twisted out of the dish until the edges are aligned.

Fix the mounting brackets to the chosen position with Rawlbolts, coachscrews or number 12 wood screws into good-quality nylon wall plugs or solid timber as is appropriate. Wind gusts can greatly increase the forces acting upon the dish aerial and mounting bracket and, therefore, fixings must be secure. Fix the dish on to the mounting bracket and adjust the direction of the dish so that it points at the chosen satellite. The elevation angle can be measured with a protractor and plumb bob as shown in Fig. 10.7.

DISH ALIGNMENT

To point the dish at the chosen satellite it must be tilted upwards (called the elevation angle) and then rotated to the left of a line running true south for the Astra satellite. This is called the azimuth angle. Figure 10.8 shows the position of this satellite in its stationary orbit above the equator.

ELEVATION ANGLE

The dish elevation angle, that is the amount of tilt required for any site in the UK, can be found by considering Table 10.3. This table gives the elevation and azimuth angles for the Astra satellite, which broadcasts the Sky channel, at 29 locations throughout the UK. When using this table, the location closest to the receiver site location should be chosen. The elevation angles in the north are less than those in the south because the north is further away from the equator and the position of the satellite as shown in Fig. 10.9.

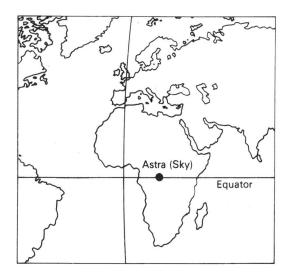

Figure 10.8 Position of the Astra satellite in its geostationary orbit above the equator.

AZIMUTH ANGLE

The dish azimuth angle is also given in Table 10.3 for various site locations.

CORRECTED AZIMUTH ANGLE

Having identified the azimuth angle for a specific satellite and site location the dish may be pointed in the correct direction with the aid of a compass. However, all magnetic compasses point to the magnetic north and magnetic south and the azimuth angles given in Table 10.3 are given for *true south*. True south is between 4° and 9° to the west of magnetic south at different parts of the UK as shown in Fig. 10.10. The azimuth angles must, therefore, be corrected for the specific locations as follows:

Find the installation site location on the map shown in Fig. 10.11 and subtract the correction values from the azimuth angles given in Table 10.3 for the Astra satellite. The azimuth angle will then be measured from the magnetic south indicated by the compass.

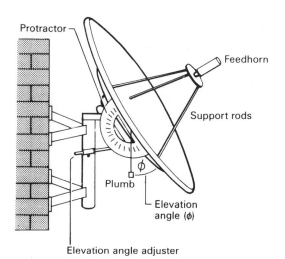

Figure 10.7 Setting up the elevation angle of a satellite dish using a protractor to measure the elevation angle ϕ.

EXAMPLE 1

A Sky dish antenna is to be installed in Manchester. The agreed place for the dish to be mounted is close to the eaves of a south-facing gable end.

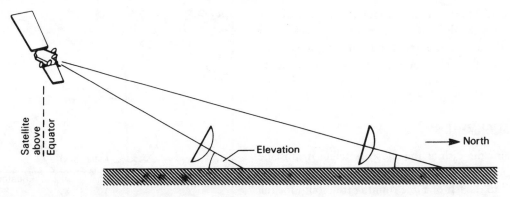

Figure 10.9 Dish elevation angle is less in the north than in the south.

From this position there is an unrestricted view of the southern horizon. Determine the elevation and azimuth angles for the location.

Table 10.3 gives an azimuth angle of 26.05° east and an elevation angle of 25.79° for the Astra satellite. Figure 10.11 gives a correction angle of 6.5° for Manchester, which must be subtracted from the azimuth angle given in Table 10.3 for the Astra satellite.

The corrected azimuth angle = 26.05° − 6.5° = 19.55°. The Sky dish must, therefore, be elevated through an angle of 25.79° measured on a protractor as shown in Fig. 10.7, and then turned left through 19.55° from the magnetic south indicated on a compass.

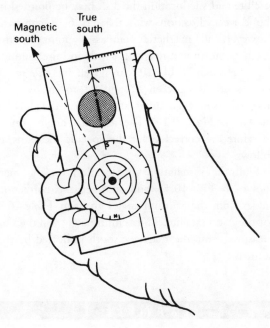

Figure 10.10 Compass readings showing true south and magnetic south.

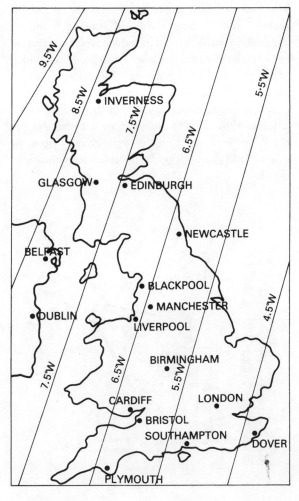

Figure 10.11 Amount of correction required to convert magnetic south to true south readings.

Table 10.3 Dish azimuth and elevation angles for various locations throughout the United Kingdom

Site locations		Satellite location (°) SUBTRACT the correction angles given in Fig. 10.11 from azimuth angles and swing dish to the LEFT of magnetic south
		Astra (Sky)
John O'Groats	Azimuth	25.61E
	Elevation	20.65
Wick	Azimuth	27.5E
	Elevation	20.85
Inverness	Azimuth	27.23E
	Elevation	21.49
Aberdeen	Azimuth	24.91E
	Elevation	22.33
Edinburgh	Azimuth	26.46E
	Elevation	23.18
Glasgow	Azimuth	27.66E
	Elevation	22.98
Newcastle upon Tyne	Azimuth	24.89E
	Elevation	24.54
Belfast	Azimuth	29.94E
	Elevation	23.66
Hull	Azimuth	23.71E
	Elevation	26.07
Manchester	Azimuth	26.05E
	Elevation	25.79
Liverpool	Azimuth	26.94E
	Elevation	25.66
Dublin	Azimuth	30.73E
	Elevation	24.7
Galway	Azimuth	33.85E
	Elevation	23.83
Great Yarmouth	Azimuth	21.57E
	Elevation	27.66
Norwich	Azimuth	22.14E
	Elevation	27.57
Birmingham	Azimuth	25.95E
	Elevation	26.85
Bedford	Azimuth	24.37E
	Elevation	27.61
Ipswich	Azimuth	21.89E
	Elevation	28.29
Cork	Azimuth	33.85E
	Elevation	25.11
Swindon	Azimuth	26.09E
	Elevation	27.81
London	Azimuth	24.1E
	Elevation	28.3
Cardiff	Azimuth	27.76E
	Elevation	27.47
Bristol	Azimuth	27.07E
	Elevation	27.68
Dover	Azimuth	22.56E
	Elevation	29.2
Southampton	Azimuth	25.82E
	Elevation	28.55
Brighton	Azimuth	24.73E
	Elevation	28.98
Plymouth	Azimuth	29.27E
	Elevation	28.17
Penzance	Azimuth	31.02E
	Elevation	28.05
Land's End	Azimuth	31.52E
	Elevation	27.93

EXAMPLE 2

A Sky dish aerial is to be installed on the south-facing wall of a London town house. The site survey has identified a suitable position. Calculate the elevation and azimuth angles for this location. Table 10.3 gives an azimuth angle of 24.1° east and an elevation angle of 28.3° for the Astra satellite. Figure 10.11 gives a correction angle of 5° for London which must be subtracted from the azimuth angle given in Table 10.3 for the Astra satellite.

The corrected azimuth angle = 24.1° − 5° = 19.1°. The Sky dish must, therefore, be elevated through an angle of 28.3° and then turned left through 19.1° from the magnetic south indicated on a compass.

Fine tuning for maximum signal strength

Having adjusted the satellite dish for the correct elevation and azimuth angles, tighten the screws sufficiently to hold the dish still and run low-loss co-axial cable to the pre-tuned receiver unit. The signal transmissions from the satellite are polarised and the feedhorn assembly must be rotated for maximum signal strength.

Connect a signal strength meter (these cost less than £100) and rotate the feedhorn and fine tune the elevation and azimuth angles for maximum signal strength. It is important to tune for maximum signal strength. On a fine day the signal from the satellite is at its strongest but rain and snow will weaken the signal which may lead to problems later, such as sparklies. Finally, tighten up all nuts and bolts while continuing to monitor the picture quality or signal strength and maintaining the best results.

PRACTICAL CONSIDERATIONS

Sparklies

Sparklies show as comet-shaped dots randomly distributed over the picture, white on dark areas and black on white areas. These are caused by a weak signal strength, and the elevation and azimuth angles should be checked if this problem is evident.

Spiders

Spiders climb inside the feedhorn, particularly in the autumn. This weakens the signal strength.

Co-axial cable

If adjustment or cleaning does not remove sparklies, then the dish itself may be too small or the co-axial down lead cable may be mismatched or too long. If the output impedance of the dish is 75 Ω, then 75 Ω impedance co-axial cable must be used. Avoid kinks or sharp bends in the co-axial down lead.

Termination

Termination of the cable at the feedhorn is best done by F-type crimp connectors. The crimp tool compresses the connector on to the jacket equally on all sides and helps to waterproof the joint.

Water ingress

Water ingress is a problem where the cable terminates at the feedhorn. Rubber boots, shrouds and heat-shrink sleeves are available and will help. The cable connector and cable should be wrapped with rubber tape at the feedhorn. PVC tape does not offer sufficient weatherproofing for this particular job because it hardens in cold weather.

Regulations

Local planning laws will allow only one dish on each building and this must be less than 90 cm in diameter. The dish must not be mounted above the roof line of the house. Installations which require bigger dishes must first obtain permission from the Local Planning Authority. An ordinary television licence will be required by satellite dish owners in addition to any monthly subscriptions to the satellite broadcasting company.

Cable television

An alternative to installing a separate roof-top receiving antenna on each house is to supply many houses with television pictures by underground cables fed from one remote ground station.

Cable companies receive satellite programmes in addition to BBC and ITV programmes and send them straight into the home via a co-axial or fibre-optic cable which plugs straight into the existing TV set.

Since private companies provide the cable TV system, the customer will pay the operator a monthly fee to view their chosen channels. At the moment it is only available in certain areas and for economic reasons will only become available to viewers in areas of high population density. However, over the past few years many thousands of miles of fibre-optic cable have been laid, making cable TV available to many more homes in the future. Many cable companies seek to achieve an economic edge by offering a combined TV, video, telephone and Internet facility down the same cable.

Computer supplies

Many computer, data processing and communication systems are sensitive to variations or distortions in the a.c. mains supply. Distortions, interference, pulses or spikes on the mains are collectively called 'noise', which can be caused by switching inductive circuits on or off, motor control equipment, brushgear, sparking of commutator motors, welders or thyristor switching of speed controllers. Variations outside very narrow limits can cause computers to 'crash' or provoke

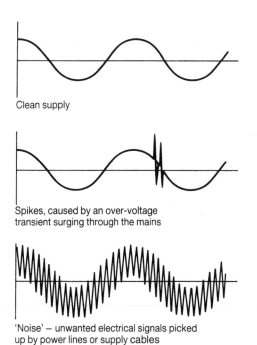

Figure 10.12 Distortions in the a.c. mains supply.

computer errors. For these reasons it is sometimes necessary to supply computer systems with a 'clean' supply or 'suppress' the possible sources of noise.

A clean supply is a separate supply which feeds items of computer equipment. This is usually fed from a point as close as possible to the LV supply source. A clean earth is also taken from the supply point, usually as one core of the feeder cable. Final distribution of the 'clean' supply is then by standard wiring circuits. Alternatively the supply to the computer equipment can be cleaned by suitable input filters. Mains suppression filters are available in many forms from any reputable supplier.

Noise generated by switch contact arcing can be suppressed by either connecting an R-C circuit across the contacts as shown in Fig. 10.13 or by connecting a metal oxide varistor (MOV) across the contacts. An MOV looks like a capacitor but behaves like a resistor, dissipating the energy contained in the noise. This has the additional benefit of increasing switch life by reducing arc damage at the contacts.

Computer networks

The electronic typewriter of the '80s has been replaced by the desktop personal computer (PC) in the office of the new millennium. If a number of personal computers are to 'talk' to each other or share access to centrally held information, records or data, then they must be connected together in a 'network'.

Local area networks of computers (LANs)

When a number of computers are connected together in the same room, office or building, they are called *local area networks* (LANs). LANs are often used in the office environment where a number of personal computers share software and central file storage facilities. A typical example of a LAN installation might be a word-processing pool where files of standard letters, customer and product details are held centrally and accessed by a number of work stations using the same word-processing software. Insurance offices and travel agents use LANs so that a number of clerical staff can access information held centrally relating to policies or holidays, issue cover notes and make bookings. LANs are also used effectively in retail outlets to maintain a record of all sales. The cash tills are microcomputers which update a central file every time an item is sold which can, in turn, update the stock control records.

TYPES OF NETWORK

LANs may be wired in different configurations to suit a particular application. A *star* network is used when a number of users require access to a central

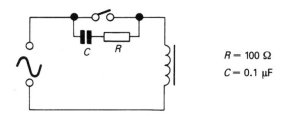

Figure 10.13 A simple noise suppressor.

$R = 100\ \Omega$
$C = 0.1\ \mu F$

store of information. The data base forms the centre point of the star, and the work stations radiate outwards.

When workstations need to communicate, for example when being used for electronic mail between offices, a *mesh* connected LAN might be used as shown in Fig. 10.14. A *bus* network is one which is widely used in offices where each workstation is connected to a central communications highway. Many network systems are commercially available. *Ethernet* is a well-known standard bus network developed by Xerox. This involves looping a 50 Ω co-axial cable around the LAN and terminating the workstations with BNC plug connectors. When workstations are removed the BNC plugs must be linked to maintain the continuity of the LAN as shown in Fig. 10.15.

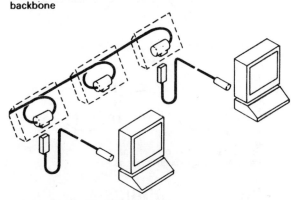

This easily installed tap for Ethernet/ IEEE 802.3 10 Base 2 network provides pluggable access to the network backbone

SIMPLE INSTALLATION
Installs quickly on IEE 802.3 10 Base 2 50-ohm cable without special tools

INTERNAL SWITCHING
Maintains series circuit so that drop cable connection and dis-connection does not disrupt operating network

Figure 10.16 LAN installation using the AMP Thinnet cable.

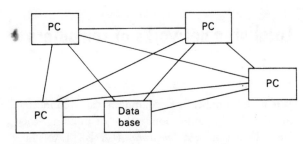

Figure 10.14 A fully interconnected network.

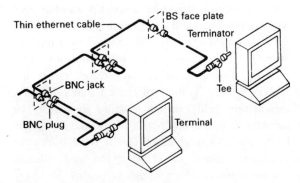

Figure 10.15 LAN cable layout.

The *Amp Thinnet* bus network is a development of the *Ethernet* network. This uses thinner RG58 (50 Ω) co-axial cable and loop in socket termination (called TAPS) which contains an internal switch so that the

LAN is not broken when a workstation is disconnected. This system can feed up to ten active workstations with a total bus length of 185 metres. The layout is shown in Fig. 10.16. The network cables are sometimes referred to as the *transmission medium* which might be co-axial cable, twisted-pair cable or flat ribbon cable. The cables can be terminated at each workstation in BNC sockets, 'D' connectors or *taps* as shown in Fig. 10.17.

Whatever the transmission medium, many cables will require containing in a suitable enclosure. Companies now invest heavily in office equipment and most are interior design conscious, recognising that the office is an important part of their public image. Making the network installation look good is, therefore, just as important as providing an efficient and flexible system. The power cables and network cables will also require segregating, not only because of the IEE Regulations, but to prevent mains-carried 'noise' on the computer network. These requirements are usually met by a multicompartment trunking system installed overhead or underfloor or run around the perimeter of the room. A skirting trunking of the type shown in Fig. 10.18 can provide an attractive solution and be fitted at floor or desktop level.

D-Connector

BNC Bayonet locking connectors

Figure 10.17 LAN network cable terminations.

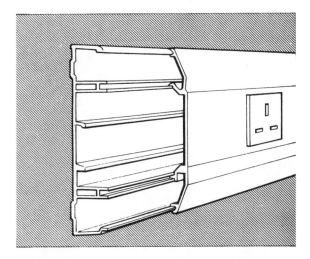

Figure 10.18 Trunking system suitable for LAN distribution which separates or segregates mains and computer cables.

11

SECURITY SYSTEMS

The installation of security alarm systems in this country is already a multi-million-pound business and yet it is also a relatively new industry. As society becomes increasingly aware of crime prevention, it is evident that the market for security systems will expand.

Not all homes are equally at risk, but all homes have something of value to a thief. Properties in cities are at highest risk, followed by homes in towns and villages and at least risk are homes in rural areas. A nearby motorway junction can, however, greatly increase the risk factor. Flats and maisonettes are the most vulnerable, with other types of property at roughly equal risk. Most intruders are young, fit and foolhardy opportunists. They ideally want to get in and away quickly but, if they can work unseen, they may take a lot of trouble to gain access to a property by, for example, removing the putty from a window.

Most intruders are looking for portable and easily saleable items such as video recorders, television sets, home computers, jewellery, cameras, silverware, money, cheque books or credit cards. The Home Office have stated that only 7% of homes are sufficiently protected against intruders although 75% of householders believe they are secure. Taking the most simple precautions will reduce the risk, installing a security system can greatly reduce the risk of a successful burglary.

Figure 11.1 Security lighting reduces crime.

Security lighting

Security lighting, such as that shown in Fig. 11.1, is the first line of defence in the fight against crime. 'Bad men all hate the light and avoid it, for fear that their practices might be shown up' (John 3:20). A recent study carried out by Middlesex University has shown that in two London boroughs the crime figures were reduced by improving the lighting levels. Police forces agree that homes which are externally well illuminated are a much less attractive target for the thief.

Security lighting installed on the outside of the home is activated by external detectors. These detectors sense the presence of a person outside the protected property and additional lighting is switched on. This will deter most potential intruders whilst also acting as courtesy lighting for visitors.

PIR detectors

Passive infra-red detector units allow a householder to switch on lighting units automatically whenever the area covered is approached by a moving body whose thermal radiation differs from the background. This type of detector is ideal for driveways or dark areas around the protected property. It also saves energy because the lamps are only switched on when someone approaches the protected area. The major contribution to security lighting comes from the 'unexpected' high-level illumination of an area when an intruder least expects it. This surprise factor often encourages the potential intruder to 'try next door'.

Passive infra-red detectors are designed to sense heat changes in the field of view dictated by the lens system. The field of view can be as wide as 180° as shown by the diagram in Fig. 11.2. Many of the 'better' detectors use a split lens system so that a number of beams have to be broken before the detector switches on the security lighting. This capability overcomes the problem of false alarms and a typical PIR is shown in Fig. 11.3.

PIR detectors are often used to switch tungsten halogen floodlights because, of all available luminaires, tungsten halogen offers instant high-level illumination. Light fittings must be installed out of reach of an intruder in order to prevent sabotage of the security lighting system.

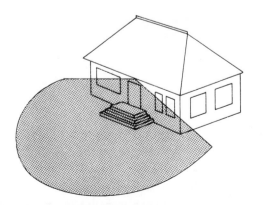

Figure 11.2 PIR detector, field of detection.

Figure 11.3 The Crabtree Minder – a typical PIR detector.

Intruder alarm systems

Alarm systems are now increasingly considered to be an essential feature of home security for all types of homes and not just property in high-risk areas. An intruder alarm system serves as a deterrent to a potential thief and often reduces home insurance premiums. In the event of a burglary they alert the occupants, neighbours and officials to a possible criminal act and generate fear and uncertainty in the mind of the intruder which encourages a more rapid departure. Intruder alarm systems can be broadly divided into three categories: those which give perimeter protection, space protection, or trap protection. A system can comprise one or a mixture of all three categories.

PERIMETER PROTECTION

A perimeter protection system places alarm sensors on all external doors and windows so that an intruder can be detected as he attempts to gain access to the protected property. This involves fitting proximity switches to all external doors and windows.

SPACE PROTECTION

A movement or heat detector placed in a room will detect the presence of anyone entering or leaving that room. Passive infra-red detectors and ultrasonic detectors give space protection. Space protection does have the disadvantage of being triggered by domestic pets but it is simpler and, therefore, cheaper to install.

Perimeter protection involves a much more extensive and, therefore, expensive installation, but is easier to live with.

TRAP PROTECTION

Trap protection places alarm sensors on internal doors and pressure pad switches under carpets on through routes between, for example, the main living area and the master bedroom. If an intruder gains access to one room he cannot move from it without triggering the alarm.

PROXIMITY SWITCHES

These are designed for the discreet protection of doors and windows. They are made from moulded plastic and are about the size of a chewing-gum packet, as shown in Fig. 11.4. One moulding contains a reed switch, the other a magnet, and when they are placed close together the magnet maintains the contacts of the reed switch in either an open or closed position. Opening the door or window separates the two mouldings and the switch is activated, triggering the alarm.

Figure 11.5 PIR intruder alarm detector.

Figure 11.6 Ultrasonic motion detector.

Figure 11.4 Proximity switches for perimeter protection.

PASSIVE INFRA-RED DETECTOR

These are activated by a moving body which is warmer than the surroundings. The PIR shown in Fig. 11.5 has a range of 12 m and a detection zone of 110° when mounted between 1.8 m and 2 m high.

ULTRASONIC DETECTOR

The ultrasonic motion detector is able to recognise the difference between random motion and intruder movement in a room. They are usually mounted in the corner of a room and have a detection range of 9 m. Figure 11.6 shows a typical ultrasonic detector.

PRESSURE PADS

Pressure pad switches, such as those shown in Fig. 11.7, are placed under the carpet close to a door. Treading on the carpet activates the switch and the alarm system.

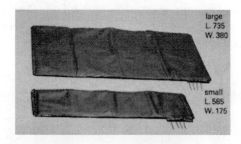

Figure 11.7 Pressure pad intruder alarm detectors.

Intruder alarm sounders

Alarm sounders give an audible warning of a possible criminal act. Bells or sirens enclosed in a waterproof enclosure, such as shown in Fig. 11.8, are suitable. The alarm sounder box now usually incorporates a

Figure 11.8 Intruder alarm sounder.

flashing light as a further deterrent. It is usual to connect two sounders on an intruder alarm installation, one inside to make the intruder apprehensive and anxious, hopefully encouraging a rapid departure from the premises, and one outside. The outside sounder should be displayed prominently since the installation of an alarm system is thought to deter the casual intruder and a ringing alarm encourages neighbours and officials to investigate a possible criminal act.

CONTROL PANEL

The control panel such as that shown in Fig. 11.9 is at the centre of the intruder alarm system. All external sensors and warning devices radiate from the control panel. The system is switched on or off at the control panel using a switch or coded buttons. To avoid triggering the alarm as you enter or leave the premises, there are exit and entry delay times to allow movement between the control panel and the door.

SUPPLY

The supply to the intruder alarm system must be secure and this is usually achieved by an a.c. mains supply and battery back-up. Nickel-cadmium rechargeable cells are usually mounted in the sounder housing box and the control panel.

Design considerations

It is estimated that there is now a one-in-twenty chance of being burgled but the installation of a security system does deter a potential intruder. Every home in this country will almost certainly contain electrical goods, money or objects of value to an intruder. Installing an intruder alarm system tells the potential intruder that you intend to make his job difficult, which in most cases encourages him to look for easier pickings.

The type and extent of the intruder alarm installation and, therefore, the cost will depend upon many factors including the type and position of the building, the contents of the building, the insurance risk involved and the peace of mind offered by an alarm system to the owner or occupier of the building.

The designer must ensure that an intruder cannot sabotage the alarm system by cutting the wires or pulling the alarm box from the wall. Most systems will trigger if the wires are cut and sounders should be mounted in any easy-to-see but difficult-to-reach position.

Intruder alarm circuits are Band I circuits and should, therefore, be segregated from mains supply cables, which are designated as Band II circuits, or insulated to the highest voltage present if run in a common enclosure with Band II cables (IEE Regulations).

Closed circuit television (CCTV)

Closed circuit television is now an integral part of many security systems. CCTV systems range from a single monitor with just one camera dedicated to

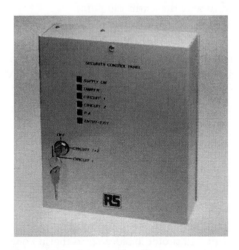

Figure 11.9 Intruder alarm control panel.

monitoring perhaps a hotel car park, through to systems with many internal and external cameras connected to several locations for monitoring perhaps a shopping precinct.

CCTV cameras are often required to operate in total darkness when floodlighting is impractical. This is possible by using infra-red lighting which renders the scene under observation visible to the camera while to the human eye it appears to be in total darkness.

Cameras may be fixed or movable under remote control, such as those used for motorway traffic monitoring. Typically an external camera would be enclosed in a weatherproof housing such as those shown in Fig. 11.10. Using remote control, the camera can be panned, tilted or focused and have its viewing screen washed and wiped.

Pictures from several cameras can be multiplexed on to a single co-axial video cable, together with all the signals required for the remote control of the camera.

A permanent record of the CCTV pictures can be stored and replayed by incorporating a video tape recorder into the system as is the practice in most banks and building societies.

Security cameras should be robustly fixed and cable runs designed so that they cannot be sabotaged by a potential intruder.

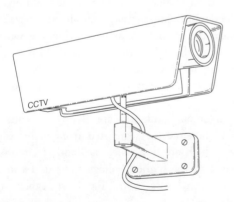

Figure 11.10 CCTV camera.

Fire security systems

A fire occurring in a building can kill the occupants by asphyxiation, poisoning or burning and, therefore, legislation has been introduced to define the premises to be protected and the standards of fire alarm equipment that should be used. The Fire Precautions Act 1971 and the Health and Safety at Work Act 1974 lay down the fundamental requirements of the legislation and BS5839 1980 *Fire Detection and Alarm Systems in Buildings* and BS3II6 1974 *Automatic Fire Alarm Systems in Buildings* set out the recommendations for the installation and servicing of fire detection and alarm systems.

Through one or more of the various statutory Acts, all public buildings are required to provide an effective means of giving a warning of fire so that life and property may be protected. An effective system is one which gives a warning of fire while sufficient time remains for any occupants to leave the building and the fire to be put out.

Fire alarm circuits are wired as either normally open or normally closed. In a *normally open circuit*, the alarm call points are connected in parallel with each other so that when any alarm point is initiated, the circuit is completed and the sounder gives a warning of fire. The arrangement is shown in Fig. 11.11. It is essential for some parts of the wiring system to continue operating when attacked by fire. For this reason the master control and sounders should be wired in M1 or FP200 cable. The alarm call points of a normally open system must also be wired in M1 or FP200 cable, unless a monitored system is used. In its simplest form, a monitored system requires a high-value resistor to be connected across the call point contacts, which permits a small current to circulate and operate an indicator declaring the circuit healthy. With a monitored system, PVC insulated cables may be used to wire the alarm call points.

In a *normally closed circuit*, the alarm call points are connected in series to normally closed contacts as shown in Fig. 11.12. When the alarm is initiated, or if a break occurs in the wiring, the alarm is activated. The sounders and master control unit must be wired in M1 or FP200 cable, but the call points may be wired in PVC insulated cable since this circuit will always be 'fail safe'.

ALARM CALL POINTS

Manually operated alarm call points should be provided in all parts of a building where people may be present and should be located so that no one need walk more than 30 m from any position within the

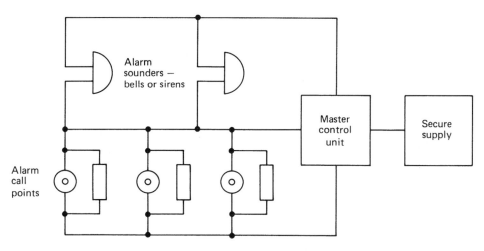

Figure 11.11 A simple normally open fire alarm circuit.

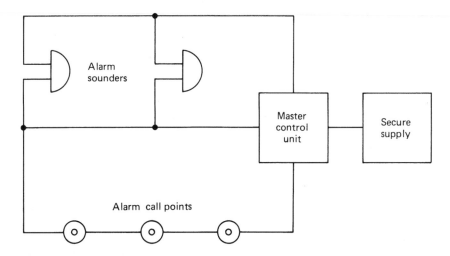

Figure 11.12 A simple normally closed fire alarm circuit.

premises in order to give an alarm. A breakglass manual call point is shown in Fig. 11.13. They should be located on exit routes and, in particular, on the floor landings of staircases and exits to the street. They should be fixed at a height of 1.4 m above the floor at easily accessible, well illuminated and conspicuous positions.

AUTOMATIC FIRE SENSORS

Automatic detection of fire is possible with heat and smoke detectors. Smoke detectors tend to give a faster response than heat detectors but a decision to use either heat or smoke detectors should be determined by their suitability for the particular installation. They should be able to discriminate between a fire and the normal environment in which they are to be installed.

Figure 11.14 shows a typical commercial heat detector and Fig. 11.15 a commercial smoke detector, both of which would be linked by mains electricity cable back to the building's fire protection system. Manual call points are suitable for occupied buildings, but in unoccupied buildings or out of hours and through the night an automatic detection system is required to protect the building.

SMOKE ALARMS

A smoke alarm is a self-contained smoke detector working on the same principle as the commercial

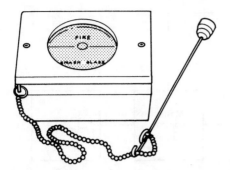

Figure 11.13 Breakglass manual call point.

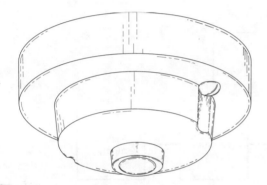

Figure 11.14 Chloride Gent heat detector.

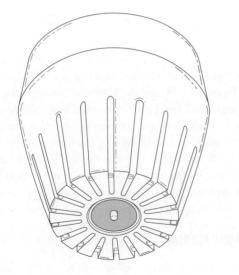

Figure 11.15 Chloride Gent smoke detector.

smoke detector. They have become very popular for installation in domestic properties because they are inexpensive, easy to install and are self-contained battery powered units.

Smoke alarms have significantly helped to reduce the number of domestic fire fatalities in countries where they are widely installed. They are an early warning device designed to sense the presence of smoke and fumes before a serious fire develops. However, make sure that the smoke alarm which you install conforms to BS5446 Part 1 and carries the British Standard Kite mark, such as that shown in Fig. 11.16.

Position of smoke alarms

Sufficient smoke must enter the smoke alarm before it will respond. The smoke alarm needs to be installed within about 10 paces or 7 metres of the potential source of fire in order to respond quickly. The alarm must also be in a position where it will be heard by everyone, even when sleeping, so that everyone can escape. A single smoke alarm will give source protection and is better than no protection but most homes should have smoke alarms fitted in all rooms where fire might break out. A small single storey home, such as a bungalow or mobile home, should have a single smoke alarm fitted in the corridor or hallway between the sleeping and living area. In dwellings with more than one sleeping area smoke alarms should be placed between each sleeping area and the living area. Dwellings on more than one level should have a smoke alarm in the downstairs hallway and a second one in the upstairs hallway. For maximum protection smoke alarms should be fitted in all rooms where fire might break out. The living room is the most likely place for a fire to break out at night, followed by the kitchen and then the dining room. Smoke alarms should also be placed in rooms where the occupants may be unable to respond effectively to a fire starting, such as the very young or the very old.

Hot smoke rises and spreads out, so the centre of the ceiling of the room is the best position for a smoke alarm. Avoid the corners of rooms where the air is dead. Fit the unit at least 30 cm (12 inches) from corners and light fittings. If a ceiling fixing position is not possible, the unit may be fitted between 15 and 30 cm (6 to 12 inches) below the ceiling on the wall.

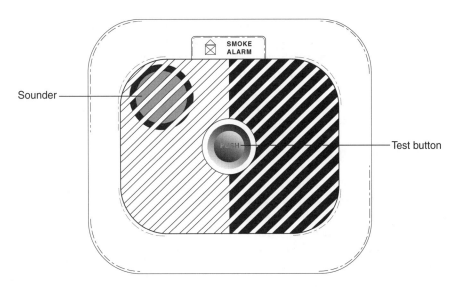

Figure 11.16 Typical domestic self contained smoke alarm.

Avoid fitting smoke alarms in bathrooms, showers, garages, attics or the peak of an A shaped ceiling because dead air at the top of these areas may prevent smoke reaching the unit. Because of the possible change in direction of airflow, smoke alarms should not be fitted in very dusty areas, next to or above heaters, air conditioning vents, windows or poorly ventilated kitchens. The kitchen is a hazardous area and special smoke alarms are available which incorporate a silence button so that the alarm can be overridden for about eight minutes if the unit is triggered by burning toast or a boiling pan.

Smoke alarm maintenance

It is recommended that domestic self-contained smoke alarms be tested every week. Pressing the test button simulates the effect of smoke so there is no reason to test the alarm with smoke or heat. Press and hold the test button until the alarm sounds, which may take up to 10 seconds. The alarm will stop sounding shortly after the button is released.

The battery should be replaced at least once per year or if it fails one of the maintenance checks described above. Low battery power is indicated by the unit beeping for a short period during the last month of battery life. It should then be replaced with a new one.

Clean the smoke alarm regularly, using a soft brush or the brush attachment of a vacuum cleaner to remove dust from the sides and cover slits where smoke enters the unit. Keep the unit closed during cleaning and do not vacuum or clean inside the unit because this might cause mechanical damage.

Limitations of smoke alarms

Smoke alarms may not wake a person who has taken drugs or alcohol. Smoke alarms are ineffective in protecting children playing with matches or protecting anyone from gas explosions or arson. They will not work if the battery is depleted or if the wrong type of battery is fitted.

There is probably only one thing worse than not installing smoke alarms, and that is the removal of the battery from an active smoke alarm for some other use or because of the nuisance of false alarms. Other occupants of the household may assume that they are being protected but the unit will be dead because it has no battery power. If frequent false alarms occur, the unit must be relocated to a more suitable position.

Smoke alarms correctly installed and maintained give excellent value for money and an early warning system in the event of fire occurring. However, they do not last indefinitely, and the manufacturers recommend that they are replaced every 10 years as a precaution against malfunction.

SOUNDERS

In a commercial fire security system the position and number of sounders should be such that the alarm can be distinctly heard above the background noise in every part of the premises. The sounders should produce a minimum of 65 dB, or 5 dB above any ambient sound which might persist for more than 30 seconds. Bells, hooters or sirens may be used but in any one installation they must all be of the same type. Examples of sounders are shown in Fig. 11.17.

FIRE ALARM DESIGN CONSIDERATIONS

Since all fire alarm installations must comply with the relevant statutory regulations, good practice recommends that contact be made with the local fire prevention officer at the design stage in order to identify any particular local regulations and obtain the necessary certification.

Larger buildings must be divided into zones so that the location of the fire can be quickly identified by the emergency services. The zones can be indicated on an indicator board situated, for example, in a supervisor's office or the main reception area. In selecting the zones, the following rules must be considered:

- Each zone should not have a floor area in excess of 2000 m².
- Each zone should be confined to one storey, except where the total floor area of the building does not exceed 300 m².
- Staircases and very small buildings should be treated as one zone.
- Each zone should be a single fire compartment. This means that the walls, ceilings and floors are capable of containing the smoke and fire.

At least one fire alarm sounder will be required in each zone, but all sounders in the building must operate when the alarm is activated. The main sounders may be silenced by an authorised person once the general public have been evacuated from the building, but the current must be diverted to a supervisory buzzer which cannot be silenced until the system has been restored to its normal operational state. A fire alarm installation may be linked to the local fire brigade's control room by the British Telecom network if the permission of the fire authority and local BT office is obtained.

The electricity supply to the fire alarm installation must be secure in the most serious conditions. In practice, the most reliable supply is the mains supply, backed up by a 'standby' battery supply in case of mains failure. The supply should be exclusive to the fire alarm installation, fed from a separate switch fuse, painted red and labelled Fire Alarm – Do Not Switch Off. Standby battery supplies should be capable of maintaining the system in full normal operation for at least 24 hours and, at the end of that time, be capable of sounding the alarm for at least 30 minutes.

Fire alarm circuits are Band I circuits and consequently cables forming part of a fire alarm installation must be physically segregated from all other circuits and from each other unless wired in MI cables (IEE Regulations 528–01).

EMERGENCY LIGHTING (BS 5266:1975)

Since an emergency occurring in a building may cause the mains supply to fail, emergency lighting should be supplied from a source which is independent from the mains supply. In most premises the alternative power supply would be from batteries, but generators may also be used. Generators can have a large capacity and duration, but a major disadvantage is the delay time

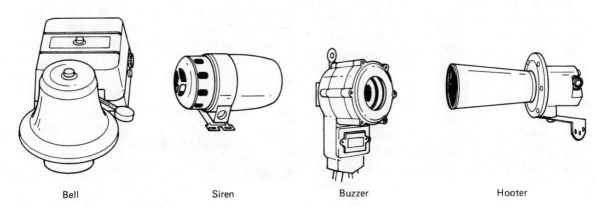

Bell   Siren   Buzzer   Hooter

Figure 11.17 Typical fire alarm sounders.

while the generator runs up to speed and takes over the load. In some premises a delay of more than five seconds is considered unacceptable and in these cases a battery supply is required to supply the load until the generator can take over.

The emergency lighting supply must have an adequate capacity and rating for the specified duration of time (IEE Regulation 313–02). BS 5266 states that after a battery is discharged by being called into operation for its specified duration of time, it should be capable of once again operating for the specified duration of time following a recharge period of not longer than 24 hours. The duration of time for which the emergency lighting should operate will be specified by a statutory authority but is normally from one to three hours. BS 5266 states that escape lighting should operate for a minimum of one hour. Standby lighting operation time will depend upon financial considerations and the importance of continuing the process or activity.

Most commercial buildings utilise self-contained emergency lighting luminaires. They incorporate energy-efficient lamps, normally supplied from the local lighting mains supply, which also trickle charges a battery pack contained within the luminaire. If the mains supply fails, the lamps are illuminated by the battery supply.

The electrician or electronics service engineer installing the emergency lighting should provide a test facility which is simple to operate and secure against unauthorised interference. The emergency lighting installation must be segregated completely from any other wiring so that a fault on the main electrical installation cannot damage the emergency lighting installation.

The batteries used for the emergency supply should be suitable for this purpose. Motor vehicle batteries are not suitable for emergency lighting applications except in the starter system of motor-driven generators. The fuel supply to a motor-driven generator should be checked. BS 5266 recommends that the full load should be carried by the emergency supply for a least one hour in every six months. After testing, the emergency system must be carefully restored to its normal operative state. A record should be kept of each item of equipment and the date of each test by a qualified or responsible person. It may be necessary to produce the record as evidence of satisfactory compliance with statutory legislation to a duly authorised person.

SENSORS AND TRANSDUCERS

A transducer is a device or element which converts one form of energy applied to the input into another form of energy at the output. The reason for wanting to do this is to convert the input signal into another form which is easier to work on, easier to amplify, easier to transmit to another place or easier to present on a display panel. It may be a device which converts mechanical vibrations into an electrical signal or which converts rotary motion into an electrical signal, as is the case with the pulse counter shown in Fig. 12.28. When used in commercial and industrial applications such as process control, the transducer generally converts a non-electrical input into an electrical output because there are many advantages to be gained by measuring non-electrical quantities by electrical and electronic instruments.

Transducers and process control

Process control and automation are concerned with handling large quantities of material whose physical and chemical properties need to be continuously monitored for quality and consistency of the finished product. Transducers convert the physical and chemical quantities into electrical signals so that the quantities can be measured. Many industrial processes create a hostile human environment and the transducer permits the quantity to be measured at a safe distance, often with much more convenience. For example, several sources of information can be observed simultaneously and compared, and the small output signals from the transducer can easily be amplified for display purposes. Extensive monitoring and control are required by industry and most systems use an appropriate transducer as the sensing element. A sensor may be defined as an electrical or mechanical device which produces an electrical signal in response to a change in the sensor's environment.

Measurement of strain

All materials deflect slightly when a force is applied to them, which causes tensile or compressive strain. Spanners bend slightly when a nut is tightened and bridges stretch and bend in high winds or when heavily loaded. A simple and convenient method of measuring this strain is to fix a *strain gauge* to the test material so that it experiences similar strains. In industry strain gauges are used during the development and testing of a product. For example, the behaviour of an aircraft wing in a wind tunnel can be monitored with strain gauges. For general engineering stress and strain analysis, a range of foil strain gauges such as those shown in Fig. 12.1 is available.

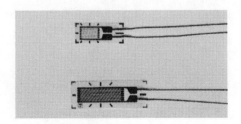

Figure 12.1 Foil strain gauges.

STRAIN GAUGES

A strain gauge is a device which experiences a change of electrical resistance when it is strained (because $R = \rho l/a\,\Omega$, as described on page 115). If the gauge is stuck firmly to the surface of a much more rigid body, any changes in dimensions of the body will cause an identical but fractional change in the dimensions of the strain gauge wire.

A modern strain gauge is formed by rolling out a thin foil of the resistive material and then cutting away parts of the foil by a photo-etching process to create the required grid pattern. This method of construction has the following advantages.

- The strain gauge is very thin.
- The gauge has a large cross-section and, therefore, a large area of contact with the test surface.
- Photo-etching techniques lend themselves to accurate reproduction and the production of 'matched sets' of gauges.
- The overall dimensions of the strain gauge can be very small so that measurements of localised strain can be made.

GAUGE FACTOR

Not all strain gauges are the same. When we calculate the strain occurring in a strain gauge we must put into the calculation a factor which reflects the 'character' of the strain gauge being used. This is called the *gauge factor* and is the constant of proportionality between the applied strain and the resistance of the gauge.

Resistance α change in dimension
$\therefore$ Resistance α strain
Resistance = $k \times$ strain
where k = the gauge factor (G.F.)
$\therefore$ Resistance = G.F. $\times$ strain

The gauge factor of most strain gauges is between 1.8 and 2.2, the total resistance varies between 60 Ω and 2000 Ω but the most common value is 120 $\Omega \pm 0.5\%$.

Active and passive axis

If the strain gauge shown in Fig. 12.2 is strained horizontally it will cause a change in the dimensions of the majority of the gauge wire and, therefore, a change in the resistance of the wire. However, if the gauge is strained vertically the loop will simply try to open out, which will have very little influence upon the dimensions of the gauge wire and, therefore, upon the resistance. The horizontal axis of the gauge is called the *active axis* and the vertical axis the *passive axis*. The change of resistance will be much greater if the gauge is strained along its active axis than if the same gauge is strained along its passive axis. Fig. 12.3 shows a single strain gauge mounted on to a specimen of metal which is being stretched by a tensile load as shown by the arrows. It is clear that the gauge mounted on metal (b) will produce the greatest change of resistance.

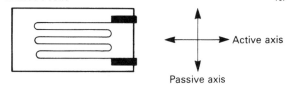

Figure 12.2 A single active axis strain gauge.

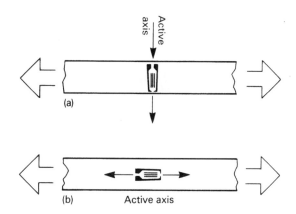

Figure 12.3 Alternative methods of mounting a strain gauge to a metal specimen.

Actual measurements

A strain gauge attached to a specimen will exhibit a change in resistance when the specimen is strained. However, the change in resistance will be very small, perhaps 1 part in 10 000, and changes of this magnitude are difficult to detect directly. They are also subject to errors due to other equally small variations such as temperature changes. To overcome these problems the strain gauge is usually connected to one arm of a Wheatstone Bridge circuit as shown in

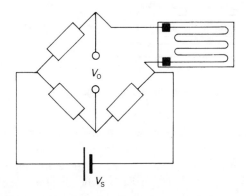

Figure 12.4 A single active gauge bridge circuit.

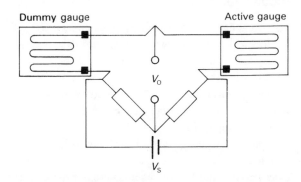

Figure 12.5 Connection of a single active gauge and dummy gauge for temperature compensation.

Fig. 12.4. The output voltage of a single active gauge bridge circuit is given by

$$V_{OUT} = \frac{V_s.G.F.e}{4} \text{ volts}$$

where V_{OUT} = the output voltage
V_S = the supply voltage
G.F. = the gauge factor
e = the strain

EXAMPLE 1

A single active strain gauge is connected to a Wheatstone Bridge circuit as shown in Fig. 12.4. Calculate the output voltage when G.F. = 2, V_s = 20 V and the strain is 0.01.

$$V_{OUT} = \frac{V_s.G.F.e}{4} \text{ volts}$$

$$\therefore V_{OUT} = \frac{20 \text{ V} \times 2 \times 0.01}{4} = 0.1 \text{ V}$$

TEMPERATURE COMPENSATION

Changes in the temperature of the specimen to which the gauge is attached will cause changes in the resistance of the gauge which are not related to the strain. These errors in strain measurement due to temperature variations can be significantly reduced by using a second *dummy gauge* to compensate for them. The dummy gauge should be mounted on an unstressed specimen of the same material and connected as another arm of the Wheatstone Bridge as shown in Fig. 12.5. The dummy gauge should be mounted as close as possible to the active gauge so that the temperature of both gauges is identical.

In practice there is some difficulty in finding an unstrained specimen of the same material close to the active gauge. This is overcome to a large extent by placing a dummy gauge on the same member but with its active axis at right-angles to the direction of strain. Foil strain gauges which have two strain elements at right-angles are available as shown in Fig. 12.6. This ensures that the two gauges are matched for resistance and temperature and are always mounted close together. Since there is still only one active gauge, the dummy gauge being unstressed, the output voltage is still given by

$$V_{OUT} = \frac{V_s.G.F.e}{4} \text{ volts}$$

TO MEASURE BENDING STRAIN

Two *active gauges* may be attached to a member as shown in Fig. 12.7 to measure the bending strain of a member. If they are mounted one on either side of

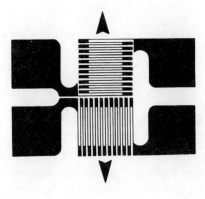

Figure 12.6 A pair of 'matched' strain gauges used for temperature compensation.

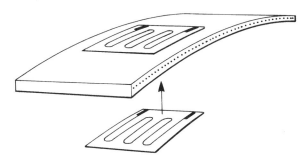

Figure 12.7 Two active gauges attached to a member measuring bending strain.

the member with their active axis along the length of the member, then a tensile strain will be imposed on the upper gauge and a compressive strain on the other. Both strain gauges will experience a resistance change of the same value but the upper gauge will increase while the lower one will decrease. Connecting them into a bridge circuit as shown in Fig. 12.8 will produce an output voltage which is twice that of a single active gauge. The output voltage for a bridge circuit which incorporates two active gauges is given by

$$V_{OUT} = \frac{V_s.G.F.e}{2} \text{ volts}$$

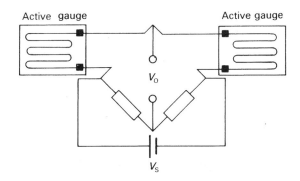

Figure 12.8 Two active gauges connected to a bridge circuit for measuring bending strain.

EXAMPLE 2

Two gauges are firmly bonded to a metal specimen to measure bending strain as shown in Fig. 12.7 Both gauges have their active axis along the length of the specimen. The input voltage is 20 V, the gauge factor 2 and the strain 0.01. Calculate the output voltage.

$$V_{OUT} = \frac{V_s.G.F.e}{2} \text{ volts}$$

$$\therefore V_{OUT} = \frac{20 \times 2 \times 0.01}{2} = 0.2 \text{ V}$$

It can be seen that the output is twice that of Example 1 which had only one active gauge. However, temperature compensation will not occur with this arrangement unless the resistors in Fig. 12.8 are replaced with dummy gauges.

TO MEASURE TENSILE STRAIN

Two active gauges can be used in a bridge circuit to measure the tensile strain in a member under tension. The gauges should be attached to the specimen as before and shown in Fig. 12.7 and connected into a bridge circuit as shown in Fig. 12.9. In this case the gauges are connected into opposite arms of the bridge and the output is once more given by the formula $V_{OUT} = V_s.G.F.e/2$ V because these are two active gauges.

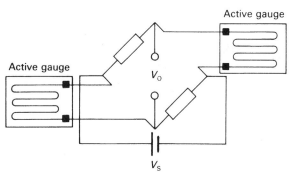

Figure 12.9 Two active gauges connected to a bridge circuit for measuring tensile strain.

Measuring the output voltage

In practice the output voltage from the bridge circuit is pretty small, which means that we can either use a sensitive voltmeter or amplify the bridge output in order to use a less sensitive voltmeter. The second option is most often preferred because it is cheaper. The 741 operational amplifier which we have already considered in Chapter 5 makes an excellent strain gauge amplifier. This is because the output voltage from the bridge occurs as the voltage difference between two points which can be connected to the inverting and non-inverting input of the op amp as shown in Fig. 12.10.

EXAMPLE 3

Two active gauges are bonded to a metal specimen held in tensile strain and connected to the circuit shown in Fig. 12.10. The gauge factor is 2 and the gain of the op amp is 100. Calculate the tensile strain on the specimen when the voltmeter is reading 2 V.

Because there are two active gauges the appropriate formula is

$$V_{OUT} = \frac{V_s.G.F.e}{2} \text{ volts}$$

Transposing this formula for strain we have

$$e = \frac{2.V_{OUT}}{V_s.G.F.} \text{ (strain has no units)}$$

V_{OUT} is the output from the bridge which is the input to the op amp. The op amp gain is 100 and, therefore, the input is 100 times less than 2 V i.e. 0.02 V.

$$\therefore V_{OUT} = 0.02 \text{ V}$$

V_s is the supply to the bridge which in this case is the voltage difference between +9 V and −9 V i.e. 18 V.

$$\therefore V_s = 18 \text{ V}$$

Substituting these values into the formula we have

$$e = \frac{2 \times 0.02 \text{ V}}{18 \text{ V} \times 2} = 0.001$$

Therefore the tensile strain on this specimen is 0.001.

Strain gauge amplifiers such as that shown in Fig. 12.11 are available commercially. The one shown encapsulates the amplifier in a 24 pin DIL package and can be used to interface any transducer with a resistive bridge configuration.

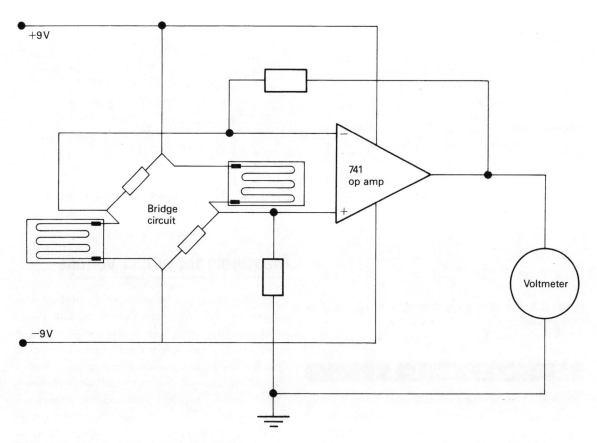

Figure 12.10 Using an op amp as a strain gauge amplifier.

Figure 12.11 A commercially available strain gauge amplifier.

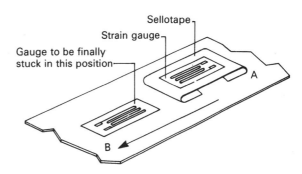

Figure 12.12 Bonding a strain gauge to the metal specimen.

Bonding the strain gauge

Probably the weakest part of a strain gauge measurement system is the adhesion of the gauge to the metal under test. The electrical connections can be tested and remade if necessary but there is no way to determine if a strain gauge is only partly bonded to the metal and giving false readings. Therefore, bonding procedures must be scrupulously followed if the strains experienced by the metal under test are to be transmitted accurately to the gauge.

SPECIMEN SURFACE PREPARATION

An area larger than the strain gauge should be cleared of all paint, rust etc., and finally smoothed with a fine-grade emery paper or fine sand blasting to provide a sound bonding surface. The area should be decreased with a solvent and finally neutralised with a weak detergent solution. Tissues or lint-free cloth should be used for this operation, wetting the surface and wiping off with clean tissues or cloth until the final tissue used is stain free. Care must be taken not to wipe grease from a surrounding area on to the prepared area or to touch the surface with the fingers. This final cleaning should take place immediately prior to the installation of the gauge.

STRAIN GAUGE PREPARATION

By sticking a short length of sellotape over the strain gauge, it can be lifted from a flat surface, taking care not to bend the gauge sharply. Holding both ends of the tape, orientate the gauge on the prepared site and stick down the end of the sellotape furthest from the connection tags. Bend the other end of the tape back upon itself to expose the underside of the gauge as shown in Fig. 12.12.

ADHESIVES

Two alternative types of adhesive are available, quick-set epoxy and cyanoacrylate. When using epoxy adhesive, apply a smooth thin coat to the whole underside of the strain gauge. Unstick end A of the sellotape, roll the gauge over in the direction of arrow B and press it down firmly into position, wiping the excess adhesive to the outside edges. Care should be taken to leave an even layer of adhesive with no air bubbles left under the strain gauge. Cover the gauge and apply a light weight or clamp until the adhesive sets. Remove the sellotape by slowly and very carefully pulling it back over itself, starting at the end furthest from the connection tags. Do not pull upwards.

If cyanoacrylate adhesive is used, stick one end of the tape down to the specimen, completely up to the gauge. Drop a small amount of adhesive in the 'hinge' point formed by the gauge and the specimen. Starting at the fixed end, with one finger push the gauge down, at the same time pushing the adhesive along the gauge in a single wiping motion until the whole gauge is stuck down. Apply pressure with one finger over the whole length of the gauge for approximately one minute. Leave for a further three minutes before removing the sellotape.

Cyanoacrylate adhesives must be handled with care because they can also bond fingers or eyelids together. Accidental contact with the adhesive should be washed away with water immediately. Accidentally bonded fingers can be peeled apart by using soap, hot water and a blunt parting tool such as the handle of a spoon.

WIRING

The lead out wires should be soldered and insulated with heat shrink sleeve or similar. The lead out wires are fragile and should be handled with care.

PROTECTION

After bonding the strain gauge and making the electrical connections, the gauge can be protected against humidity and moisture with a coat of air-drying varnish. For a more permanent installation, the gauge and its electrical connections can be encapsulated by spreading a layer of silicone rubber compound over them with a spatula, so that they become embedded in a flexible coating which dries in about 24 hours.

Measurement of pressure

Pressure is a measure of the force exerted per unit of cross-sectional area. Pressure can, therefore, be measured by monitoring the force exerted on a thin steel diaphragm which flexes under pressure as shown in Fig. 12.13. The displacement X of the diaphragm is proportional to the pressure. If a strain gauge is attached to the diaphragm a variety of indicating and recording instruments can be driven so that the pressure can be measured at some distance from the pressure transducer.

Using stainless steel as a diaphragm material results in a rugged pressure gauge able to withstand accidental overload. This type of construction is easy to waterproof and can be beam welded to withstand a variety of corrosive media. However, they are not available in diameters of less than 3.2 mm but small diameters can be achieved if the actual diaphragm is made from a silicon chip which incorporates a strain gauge. This technique permits pressure gauge diaphragms to be manufactured which have active diameters as small as 0.75 mm. The major disadvantage with silicon diaphragms is the difficulty in providing a seal and the tendency of the brittle silicon crystal to crack or shatter if overloaded.

Piezoelectric pressure transducers

An alternative to using a strain gauge as the pressure-sensing element is to use a piezoelectric crystal. If a cube of quartz crystal is squeezed across two faces, an electrical charge proportional to the pressure is generated across the faces at right-angles to the pressure as shown in Fig. 12.14. The voltage is generated directly as a result of the squeezing effect of the applied pressure. Quartz crystals are, therefore, called piezoelectric crystals because the Greek word *piezein* means to squeeze. The piezoelectric effect was discovered in 1880 by the brothers Pierre and Jaques Curie (Pierre was the husband of Marie Curie, the Nobel Prize winner). They also discovered that the reverse effect is true, that is, a dimensional change takes place when an electrical potential is applied to a quartz crystal.

In the piezoelectric pressure transducer, the crystal is contained in a steel housing and held in place by a

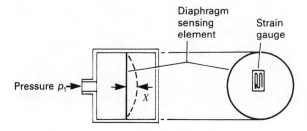

Figure 12.13 Construction of a pressure transducer.

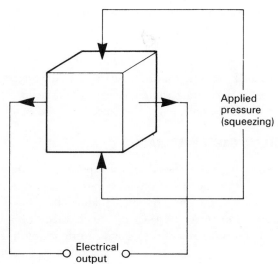

Figure 12.14 Piezoelectric effect of a quartz crystal.

preloaded sleeve, which is a thin metal cylinder as shown in Fig. 12.15. The pressure is applied to a thin steel diaphragm which deflects and converts the pressure into a compressive force on the quartz crystal. This produces a voltage at the output terminals which is converted into a usable output voltage signal by an operational amplifier, as shown in Fig. 12.16.

A piezoelectric transducer will only respond to *changes* in applied pressure and, therefore, this transducer is only suitable for dynamic pressure measurements. There is no output when a steady pressure is applied.

Bourdon tube pressure transducer

The principle of the Bourdon tube pressure gauge as a test instrument has already been described in Chapter 6. The Bourdon tube pressure gauge is a rugged, reliable, trouble-free instrument in almost universal use. The only disadvantage is that because it is entirely mechanical, it cannot be used for remote reading and it cannot produce a signal which can be used in an automatic pressure control system. However, the transducer itself, the Bourdon tube, need not necessarily be followed by mechanical signal conditioning as is the case with the Bourdon tube pressure gauge. The output from the transducer can be converted into an electrical signal if the end of the Bourdon tube is linked to the wiper arm of a variable resistor as shown in Fig. 12.17. When pressure is admitted to the Bourdon tube, the tube straightens out and causes the wiper arm to move over the resistance element. This varies the resistance in the electrical circuit which, if it is supplied by a stabilised voltage supply, will vary the current in the circuit. The ammeter scale can be calibrated in pressure units for direct reading of pressure at some distance from the transducer.

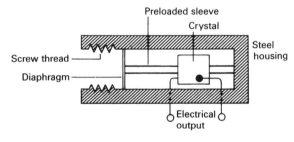

Figure 12.15 Construction of a piezoelectric pressure transducer.

Microphone pressure transducer

Sound is transmitted by pressure waves in air and, therefore, a microphone can be used as a noise-measuring transducer since it will convert a pressure wave into an electrical signal.

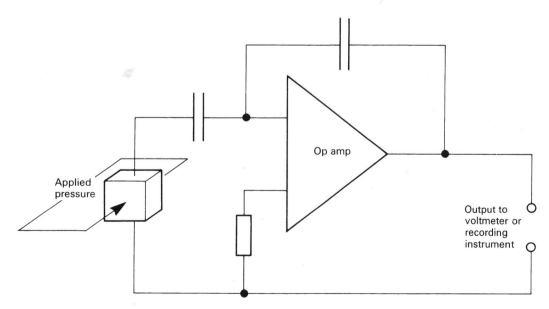

Figure 12.16 OpAmp used with a piezoelectric transducer.

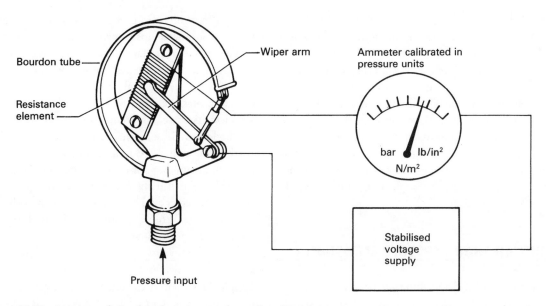

Figure 12.17 Construction of a Bourdon tube pressure transducer with an electrical output.

When a pressure wave strikes the diaphragm of the moving coil microphone shown in Fig. 12.18 the diaphragm is deflected slightly. A delicate coil wound on to the centre of the diaphragm is also deflected in the magnetic field of the permanent magnet. When any coil moves in a magnetic field, a voltage is induced in the coil. As the diaphragm vibrates in and out in response to the pressure wave fronts, the coil moves in the magnetic field and, therefore, a voltage in induced in the coil which is proportional to the sound.

A piezoelectric microphone uses a crystal of piezoelectric material in place of the moving coil. The diaphragm squeezes the mechanical axis of the crystal which produces an electrical output along the electrical axis.

A *loudspeaker* is a transducer which converts an electrical signal into a pressure wave. The construction of a loudspeaker is essentially the same as a microphone except that the diaphragm is larger. Passing a current through the moving coil of Fig. 12.18 will produce a magnetic field which will react with the permanent magnetic field causing the diaphragm to vibrate. The vibrating diaphragm produces pressure wave fronts which the human ear can identify as sound.

Measurement of temperature

Heat is a form of energy because it is capable of doing work. Temperature is a measure of the degree of hotness of an object. Temperature can be measured and its hotness indicated by a thermometer, but a simple thermometer cannot provide an indication of the temperature at a position which is remote from the sample point. For a remote indication or for the temperature measurement to form part of an automatic temperature control system we need a temperature transducer to convert the heat energy into an electrical signal.

Thermocouple temperature transducer

A thermocouple temperature transducer consists of a single pair of insulated wires of dissimilar metal fused together at one end called the hot junction, as shown in Fig. 12.19. The wire ends which are not joined are called the cold junction and are maintained at room

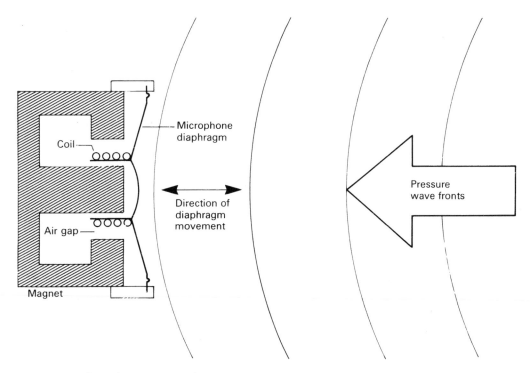

Figure 12.18 Moving coil microphone pressure transducer.

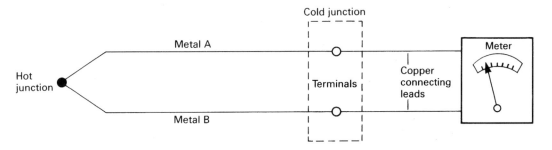

Figure 12.19 A simple thermocouple circuit.

temperature. A difference of temperature between the hot and cold junctions produces a difference of potential and, therefore, a current in milli-amperes flows. Variations of temperature produce approximately proportional variations in voltage which can be used to give an indication of the temperature. This effect was discovered in 1821 by Thomas Seebeck, a German physicist.

To identify the various wire combinations, the thermocouples are identified by the letters of the alphabet as shown in Table 12.1. Each combination of metals generates a different voltage and has a different maximum operating temperature as shown by Table 12.1. Base metal thermocouples in general generate higher voltages for the same temperature difference, but rare metal thermocouples can withstand higher maximum temperatures. For example, the maximum operating temperature of a type T thermocouple is 400°C and at this temperature it will generate a voltage of about 20 V. A type R thermocouple has a maximum operating temperature of 1600°C and at this higher temperature it generates a lower voltage, about 10 V.

However, the voltage generated by any thermocouple is very respectable and is certainly capable of easy detection and display or for operating an electro-

Table 12.1 Thermocouple types and temperature ranges

Thermocouple	*Material	Max. temperature (°C)
Base-metal		
type T	Copper–constantan	400
type E	Chromel–constantan	1000
type J	Iron–constantan	1000
type K	Chromel–alumel	1300
type N	Nicrosil/nisil	1300
Rare-metal		
type S	Platinum–platinum/10% rhodium	1600
type R	Platinum–platinum/13% rhodium	1600
type B	Platinum/30% rhodium–platinum/6% rhodium	1800

*constantan = copper/nickel; chromel = nickel/chromium; alumel = nickel/aluminium; nicrosil = nickel/chromium/silicon; nisil = nickel/silicon.

magnetic valve in a flame failure device. The actual generated voltage of different types of thermocouple at all possible temperatures are given in the BS4937.

It has become the modern practice to construct the thermocouple in a similar way to an MICC cable. The conductors are insulated with mineral insulation and are contained in a stainless steel tube varying in size from 0.25 mm to 3.0 mm diameter. Figure 12.20 shows a type T thermocouple probe with a stainless steel probe of 1.6 mm diameter and 100 mm length. This thermocouple generates approximately 42 µV/°C in the temperature range 0°C to 250°C. Therefore, over a temperature rise of 200°C, this thermocouple will generate a voltage of 200°C × 42 µV/°C = 8.4 mV.

and resistance (NTC). That is, an increase in temperature causes a decrease in the resistance of the thermistor as shown by Fig. 12.21. Since the thermistor does not generate a voltage like the thermocouple, but changes resistance in response to changes in temperature, it is normally used as one arm of the Wheatstone Bridge as shown in Fig. 12.22. As the thermistor resistance changes in response to the change in temperature the bridge becomes unbalanced and a voltage appears across the indicator.

The major advantage of the thermistor over the thermocouple is its *dramatic* change in resistance to a small change in temperature. This makes it a very *sensitive* temperature-sensing device. One major disadvantage of the thermistor is the non-linearity of the resistance change to temperature, but recently compensation networks have become available to

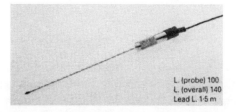

Figure 12.20 Type T hypodermic thermocouple probe.

Thermistor temperature transducer

A thermistor exhibits a very large resistance change with a change in temperature. Although some thermistors increase in resistance in response to an increase in temperature (PTC), most thermistors exhibit an inverse relationship between temperature

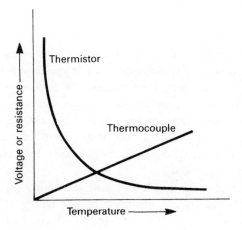

Figure 12.21 Graph showing the non-linearity of the response of a thermistor to changes in temperature.

SENSORS AND TRANSDUCERS

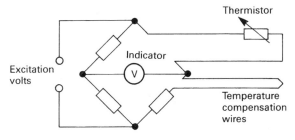

Figure 12.22 A simple thermistor circuit.

Measurement of liquid level

The measurement of liquid level in a container is an important measurement in industry and is often used in the chemical mixing process. The simplest and most frequently used level sensor is the simple float switch. A float, moving in response to the liquid's surface, operates a switch to indicate that the liquid has reached a predetermined level. The float of a modern float switch contains a small magnet which acts upon a magnetic reed switch. Fig. 12.24 shows a typical float switch. The float, which looks like a cotton bobbin, rises up the central cylinder in response to the liquid level and activates a reed switch contained in the central cylinder. The reed switch is hermetically sealed and can, therefore, operate millions of times without failure because the contacts are not exposed to atmospheric dust and corrosive fumes. The float switch shown is designed for mounting in the top of a container, but side- and bottom-mounted switches are also available. They are easily installed, reliable and require virtually no maintenance, which is an important consideration in industry.

linearise the thermistor output. It is also possible to connect the output voltage of the thermistor bridge to an analogue-to-digital converter and to use the digital output as the input to a computer. A program can then be written to linearise the thermistor output and present the temperature measurement on a suitable display.

PTC thermistors are usually used for temperature compensation and overheating protection while NTC thermistors are most often used for temperature measurement. They are manufactured from oxides of copper, manganese, nickel, cobalt and lithium, blended to give the required temperature to resistance characteristic and presented as beads, wafers or rods as shown in Fig. 12.23. They can be manufactured as extremely small devices and are, therefore, ideally suited to monitoring temperature in the most inaccessible places such as the windings of an electric motor or the core temperature of power supply cables, as well as providing current-limiting protection for electronic circuits.

When it is required to monitor a changing liquid level, the float can be made to slide up and down a long central stem containing many reed switches. The level can then be determined by observing the number of operated switches. For more exact readings of liquid level the float may be attached to an arm which operates a variable resistor, as shown in Fig. 12.25. The changing resistance indicates

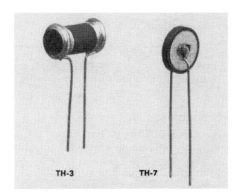

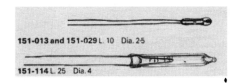

Figure 12.23 A selection of thermistors.

Figure 12.24 A float switch.

the liquid level. This type of sensor is used to measure the fuel level in the petrol tank of a motor car.

Measurement of fluid flow

Liquid flow is relatively easy to measure, and probably the most common method used by industry is the differential pressure flowmeter or head meter. When liquid passes over a curved surface or through a smaller opening than the pipe which carries it, the velocity of the liquid increases and the pressure decreases proportionately. Measuring the difference in pressure before and after the change gives an indication of the rate of flow. This principle was discovered by the Swiss mathematician Daniel Bernoulli in 1738. Knowing the rate of flow, and the cross-section of the pipe, we can calculate the flow of the liquid in gallons per minute or some other convenient unit.

Velocity flowmeters measure the rate of flow directly. The most common of these is the turbine flowmeter shown in Fig. 12.26. The liquid flow causes the turbine to rotate at a speed which is proportional to the fluid flow. Rotational speed can be sensed by a magnetic proximity switch such as that shown in Fig. 12.28 and the output displayed in suitable units.

The spinning turbine in the flowmeter shown in Fig. 12.27 breaks a photoelectric beam, giving a pulsed output which is proportional to the rate of flow. The top chamber contains an LED and a logic chip on to which is integrated all the circuitry necessary for detection, amplification and pulse shaping. The turbine motor is in direct contact with the fluid and, therefore, this type of flowmeter can only be used with clean fluids.

When mounting flowmeters, a straight length of pipe must be available on the input side of the meter to prevent problems caused by fluctuating pressures as fluids negotiate curves. The length required varies between four and fifty times the pipe diameter depending upon the type of flowmeter. A short length of straight pipe is also recommended on the discharge side.

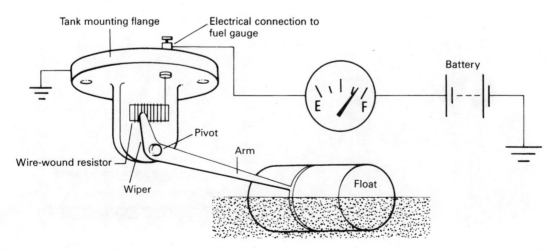

Figure 12.25 Fuel gauge float switch.

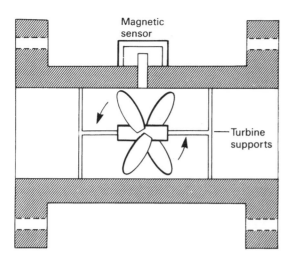

Figure 12.26 A turbine flowmeter.

L. overall = 89 mm
H. overall = 51 mm
W. overall = 37 mm

Figure 12.27 A liquid flow meter.

Measurement of speed of rotation

In most methods of measuring speed of rotation the instrument is coupled to the shaft whose speed is to be measured. The coupling may be direct with the instrument permanently connected or the instrument may be hand held to the end of the shaft while the measurement is being made.

THE TACHOMETER

This is usually a hand-held instrument on which speed is registered directly on a calibrated dial in revs/s. Tachometers are usually operated mechanically; they are useful for spot checks of speed but accuracy is not very good.

THE TACHO GENERATOR

This is a permanently coupled generator connected to the shaft end which produces a voltage proportional to the shaft speed. This allows a voltmeter, suitably calibrated in revs/s, to be placed at some distance from the machine whose speed is to be measured.

THE STROBOSCOPE

The stroboscope measures rotational speed by 'freezing' the movement so that the rotating body appears stationary. If a mark is made on the rotating shaft with chalk or whitener and a light flashed every time the mark is in the same position, the eye would only see the mark in that one position, and, therefore, the shaft would appear stationary. When taking measurements with a stroboscope the instrument is pointed at the marked shaft and the rate at which the lamp flashes is varied until the shaft appears stationary. The rate of flashing is then a measure of speed and the control dial is usually calibrated directly in revs/sec.

The advantage of this method of speed measurement is that there is no physical connection between the rotating shaft and the test equipment. The speed of very small motors found in control systems can only be measured in this way because the power required to drive a tachometer, for example, would cause the motor under investigation to slow down.

The disadvantage of this method is that we also see stationary images if the mark is illuminated once every two or three revolutions and so on. Readings are, therefore, obtained for a half, third or a quarter of the true speed. One way to avoid this error is by using some other speed-measuring device as an initial check or by calculating the approximate speed. Good practice suggests that the stroboscope should be used as follows: with the stroboscope aimed at the marked shaft, its flash rate is slowly and steadily reduced, starting with the highest rate and working down the speed range until the stationary mark is observed.

The stroboscope principle of 'freezing' movement makes this equipment useful for examining the behaviour of industrial processes which are moving too rapidly to be seen normally with the human eye. Such effects as vibrating objects, valve bounce or the turbulence of the air flow over fan blades can be studied in detail while the parts are moving at high speed.

Pulse counter

The three components of this system are a 60-toothed wheel attached to the shaft whose speed is to be measured, an electro-magnet and an electronic counter as shown in Fig. 12.28. As each tooth of the wheel approaches the end of the bar magnet the steel tooth provides a lower resistance magnetic path than did the gap between the teeth and this increases the flux density in the electromagnet and induces a voltage with a frequency equal to the number of teeth passing the electromagnet per second. The pulse counter amplifies the signal and displays it on an LED. The display can be mounted at some distance from the machine whose speed is being measured. Connection must be made by screened co-axial cables to prevent additional pulses being picked up from adjacent electrical equipment such as contactors or automatic star-delta motor starters.

Figure 12.29 shows a miniature magnetic pick-up transducer which incorporates a semiconductor IC to give a digital output every time ferrous metal passes the pole piece. The output waveform is compatible with most logic systems and is shown in Fig. 12.30. The pick-up is encased in a screwed steel cylinder 30 mm long and 6.35 mm in diameter which makes it small enough to be deployed where many conventional sensors could not be fitted.

Figure 12.29 Miniature magnetic pick-up transducer.

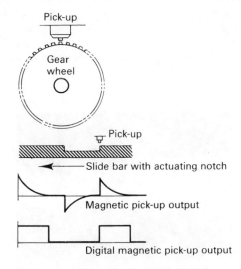

Figure 12.30 Output from a miniature magnetic pick-up.

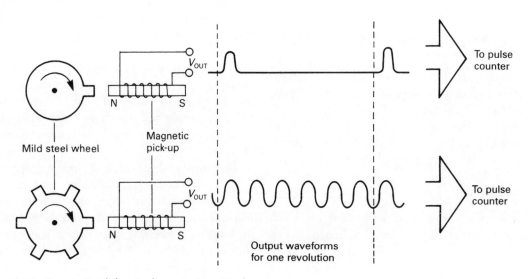

Figure 12.28 Measuring speed of rotation by counting magnetic pulses.

The Bourdon tube pressure gauge

The standard type of pressure gauge is shown in Fig. 12.31 and consists of a Bourdon tube driving a pointer through gearing across a scale. The Bourdon tube has an oval cross-section, closed at one end and bent into the arc of a circle, as shown in Fig. 12.32. When pressure is admitted, the tube tries to straighten out so that it becomes an arc of greater radius. The change is very small but the toothed quadrant and pinion mechanically amplify the displacement and the pointer moves across a suitably calibrated scale.

The principle and operation of a number of electrical measuring instruments are covered in Chapter 6 of *Advanced Electrical Installation Work*. The detailed inspection and testing procedures required by the IEE Regulations for Electrical Installations are covered in Chapter 7 of *Basic Electrical Installation Work*.

Stepper motor

A stepper motor or stepping motor is an electric motor whose shaft rotates one step at a time when the field coils are supplied in sequence by digital signals from a computer microprocessor. The number of degrees of rotation in each step is determined by the design of a particular machine. If there is 1.8° between each step, 200 steps will be required for each revolution. If there is 2.5° between each step, 144 steps will be required for each revolution. Other step angles are 3.75, 7.5, 15 and 30° giving 96, 48, 24 and 12 steps for each 360° revolution.

Each step requires an energising field coil to attract the permanent magnet rotor, as shown in Fig. 12.33. Changing the energising order of the field coils, the rotor can be made to turn forward or backwards by any number of steps. Changing the frequency of the energising pulses controls the speed of rotation.

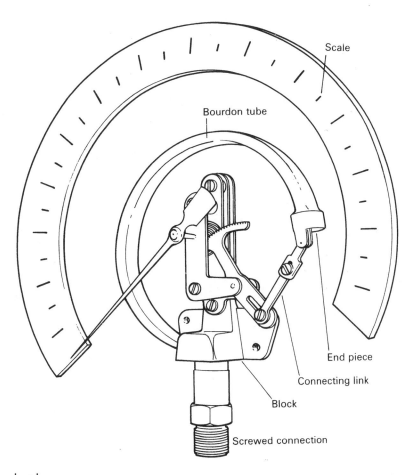

Figure 12.31 A Bourdon tube pressure gauge.

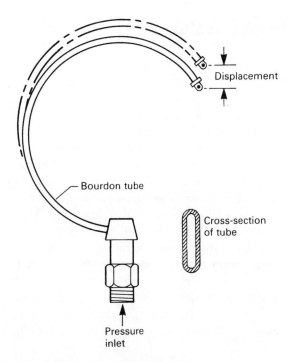

Figure 12.32 Operation of the Bourdon tube.

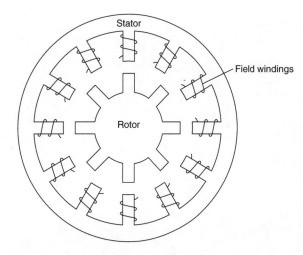

Figure 12.33 Stepper motor construction.

However, this does require a lot of 'data' and, therefore, a microprocessor generating a series of digital signals in the correct order is used to drive the motor. Figure 12.34 shows the circuit diagram of a typical control chip. When the coils of the motor are connected to this chip, it organises the logic pulses to switch the motor field coils on and off in the correct sequence.

Typical applications of stepper motors are:
- control of robot arms;
- X Y plotters;
- chart recorders;
- paper advance in printers;
- moving head drivers in disc drives;
- machine tool controls – CAD/CAM.

Characteristics of stepper motors are as follows:
- They are significantly more expensive than ordinary motors.
- They usually have four field coils which must be activated in a particular sequence and, therefore, it is essential to obtain the coil connection information before connecting the supply.
- The rotor shaft of a permanent magnet stepper motor is locked in a fixed position, even when the motor is unpowered. This is known as the *detent torque* of the motor and is due to the attraction between the permanent magnet rotor poles and the residual magnetism in the stator poles.
- The motor can be driven as slowly as is required by the application.
- The upper speed limit is governed by the maximum step rate. Two revolutions per second or 120 revolutions per minute are typical of a stepper motor.
- The motor is not damaged if the rotor shaft is stalled.
- Power consumption is high and relatively constant.
- Stepper motors normally run hot but this will not cause any damage provided that the rated voltage is not exceeded.

Industrial sensors

Many automatic industrial processes are controlled by programmable logic controllers or PLCs. A large number of input sensors are fed to the PLC which senses the logic equations programmed into the controller. The controller then activates the output devices to provide the desired outcome of the process.

For example, consider an industrial process to bake and pack biscuits. The input sensors would detect that the ingredients were mixed, that individual biscuit bases were available and that the oven was up to the desired temperature. Obtaining a logic 1 (yes) signal from all these input sensors, the PLC would

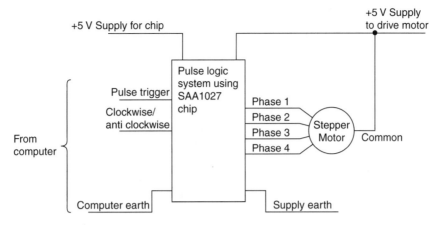

Figure 12.34 Circuit diagram of a typical microprocessor chip control to a four field coil stepper motor.

activate the output devices such as the conveyor belt to carry the biscuit bases through the oven, the weighing machine to weigh the baked biscuits and then package them in, say, 0.5 kg packets.

The input sensors might be operator switches, limit switches, photo-electric cells or proximity sensors used to sense the presence or position of a component. The output devices then act as the interface between the electronics of the PLC and the industrial machine or process. This might be an electric motor, a pump, a solenoid valve operating a gas supply or an electromagnetic relay.

Sensors can be broadly classified into two types:

- switches which generate binary (on/off) signals;
- transducers which generate analogue (variable) signals such as the Bourdon tube described earlier and shown in Fig. 12.17.

SWITCHES

Switches are dealt with in some detail in Chapter 2 of this book under the sub-headings of switches, microswitches, reed switches and electromagnetic relays. You should read those sections in conjunction with these topics.

Limit switches

In limit switches the electrical contacts of an ordinary switch are made or broken by some mechanical device such as a lever, spring or plunger. The small and sensitive mechanical switch is usually fitted with a lever or activator so that only a small force is required to operate a snap action switch. The activator may be a simple lever or incorporate a roller as shown in Fig. 12.35. The limit switch shown is available with rolling heads of various diameters and arm lengths to suit various applications.

Let us now consider a simple industrial process. The control valve shown in Fig. 12.36 controls the flow of liquid in one part of the process. Turning to the left, the fluid flow is off, turning to the right, the fluid flow is on.

With the control valve off, the 'P' limit switch is activated and sends a logic 1 (yes or on) input to the PLC. To turn the valve on, the PLC sends a supply of say 24 V to output 'Q' and observes that 'P' activates. The purpose of limit switch 'R' is to tell the PLC controller that the valve is now fully open and signal 'Q' can therefore be switched off, leaving the control valve in the open position. Without the signal from the limit switch 'R' the PLC controller would have no way of knowing if the control valve was fully open or had stuck half way open.

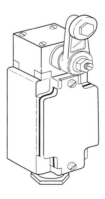

Figure 12.35 Industrial limit switch with roller lever.

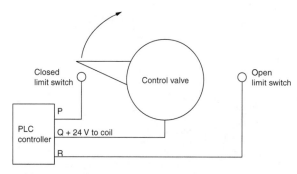

Figure 12.36 Limit switch controlling a valve.

Limit switch fixing methods – good practice

- Always make sure that: electrical connections are tight;
- any cord grip or entry gland is securely tightened but not overtightened;
- the limit switch is fixed in a position that does not kink or sharply bend the connecting flexible cable;
- a single loop is provided in cables feeding limit switches attached to vibrating machinery;
- the limit switch actuator is adjusted so that the limit switch operating plunger is not applied too harshly or overoperated in the tripped position;
- the control cable does not rub on any vibrating or moving equipment.

Proximity switches

Proximity switches are important components in any automated control system. They transmit information relating to the operating conditions of the machine to the PLC controller. Proximity switches are able to sense the close 'proximity' of a component (termed the 'target') without the target actually touching them. The gap between the switch's 'sensing face' and the target is typically 5–10 mm. Proximity switches are usually either *inductive* or *capacitive* in their mode of operation, although *retro-flective* photocells are also used (these are photocells which transmit light and sense when it is being reflected back by the presence of some reflective surface) and we will consider these separately.

Proximity switches are encapsulated as a cylinder or block as shown in Fig. 12.37. The cylinder or block encapsulates an electronic solid state oscillator and an output driver switch. The oscillator detects an object of a size which approximately matches the size of the sensing face when it is brought close to the sensing face (5–10 mm), which operates the output driver and switch.

A *capacitive proximity sensor* encapsulates an oscillator whose capacitors constitute the sensing face. When a conducting or insulating material with a permittivity greater than 1 is placed within the sensing field it modifies the coupling capacitance and starts the oscillator, as shown in Fig. 12.38. This causes the output driver to operate and a normally open or normally closed output signal is produced.

A capacitive proximity switch detects objects which have a permittivity greater than 1 and is, therefore, ideal for detecting insulating materials such as paper, cardboard and glass or liquid materials. For example:

air has a permittivity of 1
paper has a permittivity of 3.7
glass has a permittivity of 5 to 10
paraffin has a permittivity of 2.2
castor oil has a permittivity of 4.5
ethyl alcohol has a permittivity of 26
water has a permittivity of 80.37

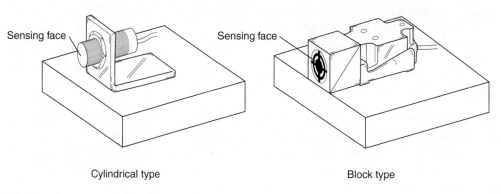

Figure 12.37 Physical appearance of industrial proximity sensors.

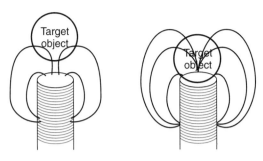

Figure 12.38 Capacitive proximity sensor – coupling capacitance modified by target object being detected.

An *inductive proximity sensor* encapsulates an oscillator whose windings constitute the sensing face. An alternating magnetic field is generated in front of the sensing face, as shown in Fig. 12.39. When a metal target enters the alternating magnetic field, eddy currents are induced in the target, which draws energy from the proximity sensor and the electromagnetic field collapses. This causes the output driver to operate and a normally open or closed output signal is produced.

The description above of the 'basic principle' suggests a crude device, but the inductive proximity sensor is very sensitive in practice. As the target object passes over the proximity switch it switches on when the target reaches any point on the inner line shown in Fig. 12.40, and turns off when it reaches the outer line. The difference between the turn on and turn off point is called hysteresis and provides stable operation by preventing the switch from chattering between the on and off states.

Proximity sensor – good practice

The proximity sensor should always operate at between 50% and 80% of its range in order to avoid

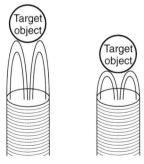

Figure 12.39 Inductive proximity sensor – magnetic field modified by target object being detected.

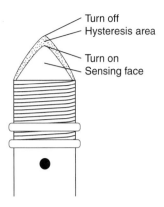

Figure 12.40 Turn on and turn off point on an inductive proximity sensor.

marginal conditions. The most obvious factor which determines sensing range is the physical size of the switch. The strength of the radiated field is a function of the core size. Large switches have large cores and, therefore, longer sensing ranges.

The range of an inductive proximity switch is specified for a mild steel target. When the target is not mild steel, a correction factor must be applied. For example, the correction factor for stainless steel is 0.7 and, therefore, the actual range for a stainless steel target will be 70% of the nominal specified range. The correction factor for aluminium is 0.4 and the range is, therefore, 40% of that specified.

The size and shape of the target object also affect the sensing range. The specified range is based on a target that is at least as large as the sensing face. Targets which are smaller or a different shape such as a toothed wheel or a screw head will result in shorter sensing ranges.

Proximity sensors may be two wire or three wire. The difference between the two types is how the electrical supply drives the device.

- *Two wire types* are wired in series with the load. They are sometimes called load powered switches. When they are open there is some leakage current because of the power required to operate the sensor. Most of the supply voltage is dropped across the switch but because of the leakage current a small voltage is applied to the load.
- *Three wire types* use two wires to energise the proximity sensor, live and neutral and the third wire is the switch output wire. When a three wire proximity switch is open there is zero leakage

current and zero volts applied to the load. All three wire proximity switches are polarity sensitive.

Always replace like proximity switches with like. Basically three types of proximity sensors are available as shown in Fig. 12.41.

- The pre-cabled type offers good resistance to moisture and is ideal for machine tool applications.
- The connector type offers ease of installation and maintenance.
- The screw terminal type offers flexibility through adaptability of the cable lengths.

When mounting sensors:

- carefully adjust the target distances to be between 50% and 80% of the specified operational range;
- ensure that the support is sufficiently rigid and large enough to resist the effects of shock and vibration;
- do not over tighten the mounting nuts on threaded tubular sensors – it is better to use an adhesive such as Locktite to secure the nuts if vibration might be a problem;
- use the screwdriver supplied with the sensor to make adjustments;
- avoid stress on the wiring by incorporating a single turn loop;
- make sure all connections are electrically secure and that the control cable does not rub on any vibrating or moving surface.

Photo-electric detectors

Photo-electric detectors are light transmitting devices which switch by either reflecting the light back to the transmitter or by interrupting the reflected light as shown in Fig. 12.42. The photo-electric detector

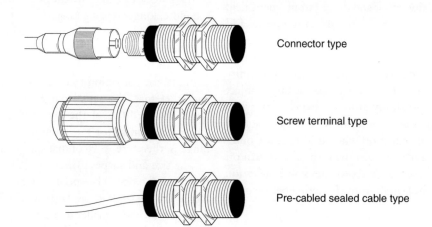

Figure 12.41 Three types of proximity sensor.

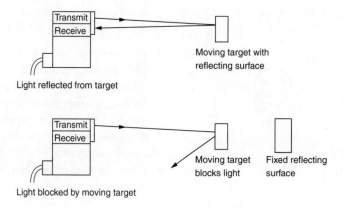

Figure 12.42 Basic principle of photo-electric detector.

incorporates an amplifier and both transmits and receives the polarised red light, giving trouble free operation and freedom from stray ambient or reflected light. Alternatively, the optical head can be fed from a remote amplifier through a fibre-optic cable or a remote optical head. The selection of the photo-electric detector is determined by the working environment. A.c. or d.c. operation is available in ranges from 24 V to 230 V.

Photo-electric detector – good practice

- Correctly position the reflector and adjust the sensitivity control on the detector until the reflector is strongly detected;
- Adjust the sensing distance so that it does not include the background;
- Check that the detector switches on when the target object is present;
- Ensure that the reflector and detector are rigidly mounted;
- Standard reflectors do not work at close range, a special reflector with large trihedrons will be required;
- Avoid electromagnetic interference by separating control wiring from mains power wiring by at least 10 cm;
- Standard photo-electric detectors are designed to have a high level of immunity to surrounding light. However, do be sensitive to the possibility of interference from sunlight when selecting the sensor position. Also, industrial environments may be dusty or dirty and, therefore, maintenance procedures must ensure sensing heads are cleaned and that seals, cable glands and blanking plugs are carefully fitted to optimise performance.

RELAYS

An electromagnetic relay is simply an electromagnet operating a number of switch contacts as shown in Fig. 12.43. When a current is passed through the coil contacts 13–14, the soft iron core becomes magnetised, attracts the iron armature and closes the switch contacts 9–5, 10–6, 11–7 and 12–8. The relay coil is electrically insulated from the switch contacts and, therefore, a relay is able to switch circuits operating at a different voltage to the coil operating voltage. The small current which energises the coil is also able to switch larger currents at the

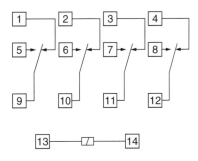

Figure 12.43 Connection diagram for a typical industrial plug in relay.

switch contacts. The switch part of the relay may have many poles controlling several circuits at once as shown by the pin connection diagram given above.

Miniature plug-in relays are popular in industrial electronic circuits. A typical example is shown in Fig. 12.44. However, all mechanical-electrical switches are limited in their speed of operation by the time taken physically for a movable contact to make or break a switch contact. Where extremely high speed operations are required, the switching action must take place without physical movement. This is only possible using the properties of semiconductor materials in devices such as transistors and thyristors. They permit extremely high speed switching without arcing and are considered in Chapter 4 of this book.

Figure 12.44 An electromagnetic relay.

Opto electronics

The word is derived from optics, the scientific study of sight and light and electronics, the use of semiconductor devices. Opto electronics covers fibre-optic cables which are used in communication systems and discussed in Chapter 10 of this book; also optical displays such as light-emitting diodes and seven-segment number and letter displays, such as those shown in Fig. 12.45. Finally, opto electronics covers

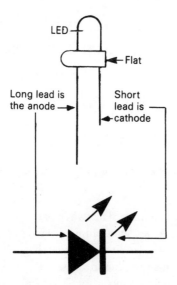
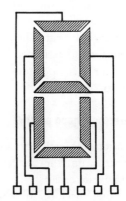

Figure 12.45 Opto electronic LED and seven segment displays.

optical sensors which are used for coupling and isolation in industrial processes.

Opto isolators provide electrical isolation using the properties of an LED to switch a transistor. The light energy from the LED switches the transistor to the on position. This arrangement can give electrical isolation up to 10 kV and a very rapid switching time of less that one micro-second.

The OPI 110 series of opto isolators are suitable for use in equipment which is designed to be intrinsically isolated for use in the very hazardous flameproof and explosive situations to be found in the chemical and petroleum industries. These opto isolators carry the prestigious EX and BASEEFA approval marks for use in hazardous areas. Opto isolators and coupling sensors are either packaged in a cylindrical arrangement or in the 6 pin DIL (dual-in-line) package shown in Fig. 12.46. Each opto electronic device has its own place in the circuit design for achieving a particular solution and must be chosen carefully from a manufacturer's catalogue or the reference numbers matched to the replacement device.

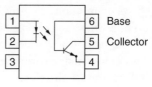

Figure 12.46 Opto isolator packaged as a 6 pin DIL configuration.

APPENDICES

Appendix A: Obtaining information and components

For local suppliers, you should consult your local telephone directory. However, the following companies distribute electrical and electronic components throughout the UK. In most cases, telephone orders received before 5 pm can be dispatched the same day.

Electromail (R. S. mail order business), P.O. Box 33, Corby, Northants NN17 9EL Tel. (011536) 204555.

Farnell Electronic Components, Canal Road, Leeds LS12 2TU Tel. 0113 636311.

Maplin Electronics, P.O. Box 777, Rayleigh, Essex SS6 8LV Tel. 01268 552961.

Rapid Electronics Ltd, Heckworth Close, Severalls Industrial Estate, Colchester, Essex CO4 4TB Tel. (01206) 751166. Fax (01206) 751188. email sales @ rapidelec.co.uk.

R.S. Components Ltd, P.O. Box 99, Corby, Northants NN17 9RS Tel. (01536) 201234.

Verospeed Electronic Components, Boyatt Wood, Eastleigh, Hants SO5 4ZY Tel. (02380) 644555.

Appendix B: Abbreviations, symbols and codes

Abbreviations used in electronics for multiples and sub-multiples

T	tera	10^{12}
G	giga	10^{9}
M	mega or meg	10^{6}
k	kilo	10^{3}
d	deci	10^{-1}
c	centi	10^{-2}
m	milli	10^{-3}
μ	micro	10^{-6}
n	nano	10^{-9}
p	pico	10^{-12}

Terms and symbols used in electronics

Term	Symbol
Approximately equal to	$\simeq$
Proportional to	$\propto$
Infinity	∞
Sum of	Σ
Greater than	$>$
Less than	$<$
Much greater than	$\gg$
Much less than	$\ll$
Base of natural logarithms	e
Common logarithms of x	$\log x$
Temperature	θ
Time constant	T
Efficiency	η
Per unit	p.u.

Electrical quantities and units

Quantity	Quantity symbol	Unit	Unit symbol
Angular velocity	ω	radian per second	rad/s
Capacitance	C	farad	F
		microfarad	µF
		picofarad	pF
Charge or quantity of electricity	Q	coulomb	C
Current	I	ampere	A
		milliampere	mA
		microampere	µA
Electromotive force	E	volt	V
Frequency	f	hertz	Hz
		kilohertz	kHz
		megahertz	MHz
Impedance	Z	ohm	Ω
Inductance, self	L	henry (plural, henrys)	H
Inductance, mutual	M	henry (plural, henrys)	H
Magnetic field strength	H	ampere per metre	A/m
Magnetic flux	θ	weber	Wb
Magnetic flux density	B	tesla	T
Potential difference	V	volt	V
		millivolt	mV
		kilovolt	kV
Power	P	watt	W
		kilowatt	kW
		megawatt	MW
Reactance	X	ohm	Ω
Resistance	R	ohm	Ω
		microhm	µΩ
		megohm	MΩ
Resistivity	ρ	ohm metre	Ωm
Wavelength	λ	metre	m
		micrometre	µm

Capacitor values-conversion table

Capacitance (picofarad pF)	Capacitance (nanofarad nF)	Capacitance (microfarad μF)	Capacitance code
10	0.01		100
15	0.015		150
47	0.047		470
82	0.082		820
100	0.1		101
330	0.33		331
470	0.47	0.00047	471
1000	1.0	0.001	102
1500	1.5	0.0015	152
2200	2.2	0.0022	222
4700	4.7	0.0047	472
6800	6.8	0.0068	682
10 000	10	0.01	103
22 000	22	0.022	223
47 000	47	0.047	473
100 000	100	0.1	104
220 000	220	0.22	224
470 000	470	0.47	474

Capacitance code: First two digits significant figures; third is number of zeros. Value given in pF.

Suffixes used with semiconductor devices

Many semiconductor devices are available with suffix letters after the part number, i.e. BC108B, C106D, TIP31C.
The suffix is used to indicate a specific parameter relevant to the device — some examples are shown below.

Thyristors, triacs, power rectifiers
Suffix indicates voltage rating, e.g. TIC 106D indicates device has a 400 V rating. Letters used are:

Q = 15 V   B = 200 V   M = 600 V
Y = 30 V   C = 300 V   S = 700 V
F = 50 V   D = 400 V   N = 800 V
A = 100 V   E = 500 V

Small signal transistors
Suffix indicates h_{FE} range, e.g. BC108C
A = h_{FE} of 125–260
B = h_{FE} of 240–500
C = h_{FE} of 450–900

Power transistors
Suffix indicates voltage, e.g. TIP32C
No suffix = 40 V
A = 60V
B = 80V
C = 100V
D = 120V

Resistor and capacitor letter and digit code (BS 1852)

Resistor values are indicated as follows:

0.47 Ω	marked	R47	100 Ω	marked	100R
1 Ω		1R0	1 kΩ		1K0
4.7 Ω		4R7	10 kΩ		10K
47 Ω		47R	10 MΩ		10M

A letter following the value shows the tolerance.
F = ± 1%; G = ±2%; J = ±5%; K = ±10%; M = ±20%;
R33M = 0.33 Ω ±20%; 6K8F = 6.8 kΩ ±1%.

Capacitor values are indicated as:

0.68 pF	marked	p68	6.8 nF	marked	6n8
6.8 pF		6p8	1000 nF		1μ0
1000 pF		1n0	6.8 μF		6μ8

Tolerance is indicated by letters as for resistors. Values up to 999 pF are marked in pF, from 1000 pF to 999 000 pF (= 999 nF) as nF (1000 pF = 1 nF) and from 1000 nF (= 1 μF) upwards as μF.

Some capacitors are marked with a code denoting the value in pF (first two figures) followed by a multiplier as a power of ten (3 denotes 10^3). Letters denote tolerance as for resistors but C = ± 0.25 pF.
e.g. 123J = 12 pF × 10^3 ± 5% = 12 000 pF (or 0.12 μF).

Appendix C: Greek symbols

Greek letters used as symbols in electronics

Greek letter	Capital (used for)	Small (used for)
Alpha	—	α (angle, temperature coefficient of resistance, current amplification factor for common-base transistor)
Beta	—	β (current amplification factor for common-emitter transistor)
Delta	Δ (increment, mesh connection)	δ (small increment)
Epsilon	—	ε (permittivity)
Eta	—	η (efficiency)
Theta	—	θ (angle, temperature)
Lambda	—	λ (wavelength)
Mu	—	μ (micro, permeability, amplification factor)
Pi	—	π (circumference/diameter)
Rho	—	ρ (resistivity)
Sigma	Σ (sum of)	σ (conductivity)
Phi	Φ (magnetic flux)	ϕ (angle, phase difference)
Psi	Ψ (electric flux)	—
Omega	Ω (ohm)	ω (solid angle, angular velocity, angular frequency)

Appendix D: Battery information

1. TYPES OF BATTERY

Alkaline primary cells

These cells and batteries offer very long service life compared with Leclanché types in equipments having high current drains. In addition these cells have very low self-discharge currents and are completely sealed.

Available in sizes AAA, AA, C, D and PP3.

Silver/mercuric oxide primary cells

These button cells are suitable for use in calculators, small tools, cameras, clocks, watches etc. They may often be used as replacements for previously fitted alkaline manganese button cell types. Supplied in boxes of individual blister packs.

Available in six of the most popular sizes.

Ni-Cad sintered cells

Applications where extreme ruggedness and/or high peak currents are required. In addition these cells offer very long service life and can be electrically misused without damage.

Available in sizes N, AAA, AA, C, D and PP9.

Ni-Cad high-temperature sintered cells

Primarily for use in emergency lighting installations these cells and batteries are particularly suitable for charging and discharging at elevated temperatures. Other applications include alarm control panels and emergency and standby areas where higher ambient temperatures are experienced.

Available as single D cells and $3 \times D$ cell battery packs.

Note: sintered cells

Have fairly high self-discharge currents and are therefore not suitable for equipment which has to be operational without recharging after being left unattended for long periods of time.

Ni-Cad mass plate cells

Applications where small size and ruggedness are required. These cells have low self-discharge currents and are ideal in small portable equipment.

A range of sizes including PP3 is available.

2. BATTERY RATING AND STORAGE

Battery type and Stock Nos	Ratings Voltage Fully charged	Discharged	Capacity	Shelf life (see note 1)	Storage Storage life	Storage temp.
Alkaline Cells						
591–657 AAA	1.50 V	0.90 V	0.7 Ah			
591–225 AA	1.50 V	0.90 V	1 Ah	24	24	
591–231 C	1.50 V	0.90 V	4 Ah	months	months	−20°C−+50°C
591–247 D	1.50 V	0.90 V	8 Ah	$T_a = 20°C$	@$T_a = 20°C$	
591–792 PP3	9.00 V	5.40 V	0.4 Ah		(see note 2)	
Silver Oxide Cells						
592–082 SR41	1.55 V	1.2 V	38 mAh	24 months	24 months	
592–098 SR43	1.55 V	1.2 V	120 mAh	$T_a = 20°C$	@$T_a = 20°C$	−10°C−+25°C
592–105 SR44	1.55 V	1.2 V	140 mAh	to 90%	to 90%	recommended
592–111 SR54	1.55 V	1.2 V	80 mAh	capacity	capacity	
Mercuric Oxide Cells						
592–127 PX/RM 625	1.35 V	0.9 V	350 mAh	24 months	24 months	
592–133 PX/RM 400	1.35 V	0.9 V	70 mAh	$T_a = 20°C$	@$T_a = 20°C$	−10°C−+25°C
				to 90%	to 90%	recommended
				capacity	capacity	
Ni-Cad Sintered Cells			(5 hr discharge rate)			
592–026 N	1.24–1.27 V	1.00 V	150 mAh			
591–146 AAA	1.24–1.27 V	1.00 V	180 mAh	120 days		
591–051 AA	1.24–1.27 V	1.00 V	500 mAh	$T_a = 0°C$		
591–045 C	1.24–1.27 V	1.00 V	2 Ah	40 days	>5 years	−40°C−+60°C
591–039 D	1.24–1.27 V	1.00 V	4 Ah	$T_a = 20°C$	(see note 2)	
591–095 PP9	8.68–8.90 V	7.00 V	1.2 Ah	11 days		
				$T_a = 40°C$		
Ni-Cad High Temp Cells			(5 hr discharge rate)			
592–032 D	1.24–1.27 V	1.00 V	4 Ah			−45°C−+65°C
592–048 3 × D, Stick	3.72–3.81 V	3.00 V	4 Ah	55 days	>5 years	possible
592–054 3 × D, Plate	3.72–3.81 V	3.00 V	4 Ah	$T_a = 20°C$	(see note 2)	0°C−+45°C recommended
Ni-Cad Mass Plate			(10 hr discharge rate)			
591–477 PCB Battery	3.72–3.81 V	3.00 V	100 mAh	26 months		Max limits
591–089 PP3	8.70–8.90 V	7.00 V	110 mAh	$T_a = 0°C$	>5 years	−40°C−+50°C
591–168 Button Cell	1.24–1.27 V	1.00 V	170 mAh	10 months	(see note 2)	0°C−+45°C
591–174 Button Cell	1.24–1.27 V	1.00 V	280 mAh	$T_a = 20°C$		recommended
591–180 Stack	8.70–8.90 V	7.00 V	170 mAh	1 month		
591–196 Stack	6.00–6.20 V	5.00 V	280 mAh	$T_a = 40°C$		

Note:
1. Period after which only 60% of the stated capacity is obtainable.
2. Period after which battery should be replaced with new stock.

3. CHARGING INFORMATION FOR RECHARGEABLE BATTERIES

Battery type and stock numbers		Charge mode	Charge rate Continuous	Max	Charge $\emptyset$ (note 1)	Temp range	Stock numbers for suitable RS chargers
Alkaline Cells							
591–657	AAA						
591–225	AA			Not rechargeable			
591–231	C						
591–247	D						
591–792	PP3						
Silver Oxide Cells							
592–082	SR41						
592–098	SR43			Not rechargeable			
592–105	SR44						
592–111	SR54						
Mercuric Oxide Cells							
592–127	PX/RM 625			Not rechargeable			
592–133	PX/RM 400						
Ni-Cad Sintered Cells							
592–026	N		15 mA	2C			
591–146	AAA		2 mA	2C	60% @ C/8		A wide range of suitable RS chargers is available. Please refer to current RS catalogue. (see note 3)
591–051	AA	Series constant current	66 mA				
591–045	C		250 mA	10C	80% @ C/2	+10°C–+45°C	
591–039	D		500 mA				
591–095	PP9		100 mA	2C	90% @ C		
Ni-Cad High Temp. Cells							
592–032	D		500 mA			(see note 2)	591–067, 591–714
592–048	3 × D, Stick	Series constant current	500 mA	2C	84%	–20°C–+65°C (reduced spec) +10°C–+35°C (full spec)	591–067
592–054	3 × D, Plate		500 mA				591–067
Ni-Cad Mass Plate							
591–477	PCB Battery		1.0 mA				–
591–089	PP3		1.1 mA			Max limits	591–152
591–168	Button Cell	Series constant current	1.7 mA	C/10	72%	0°C–+45°C	–
591–174	Button Cell		2.8 mA			+10°C–+35°C recommended	–
591–180	Stack		1.7 mA				–
591–196	Stack		2.8 mA				–

Notes:

1. $\emptyset = \dfrac{\text{energy stored in the battery}}{\text{energy supplied to the battery}}$

2. At temperatures below 0°C charge current is limited to 120 mA and voltage to a maximum of 1.55 V/cell.
3. R.S. Components Ltd. (See Appendix A for further information)

4. DISCHARGE INFORMATION

Battery type and stock numbers		Discharge temp	1 max (see note 1)	Cyclic life	Standby life	R_{int} (dc)	R_{int} (ac) 50 Hz
Alkaline Cells							(see note 2)
591–657	AAA						175 mΩ
591–225	AA						160 mΩ
591–231	C	−20°C−+50°C	0.9 A	1 cycle	2 years		130 mΩ
591–247	D						120 mΩ
591–792	PP3						2.7 Ω
Silver Oxide Cells			(see note 5)				
592–082	SR41		5/10 mA			8 Ω	
592–098	SR43		10/50 mA			6 Ω	
592–105	SR44	−10°C−+60°C	10/50 mA	1 cycle	2 years	4 Ω	
592–111	SR54		10/50 mA			10 Ω	
Mercuric Oxide Cells				(see note 5)			
592–127	PX/RM625	−10°C−+60°C	5/10 mA	1 cycle	2 years	2 Ω	
592–133	PX/RM400		0.1/0.5 mA			5 Ω	
Ni-Cad Sintered Cells							
592–026	N		0.9 A			105 mΩ	
591–146	AAA		1.0 A			80 mΩ	
591–051	AA		35 A	700–1000		20 mΩ	
591–045	C	−30°C−+50°C	70 A	cycles	7 years	8 mΩ	
591–039	D		110 A			5 mΩ	
591–095	PP9		2.4 A				
Ni-Cad High temp Cells		(see note 3)					(see note 2)
592–032	D	−40°C−+65°C				6.5 mΩ	3.75 mΩ
592–048	3 × D Stick	(reduced spec) −20°C−+45°C	24 A	800–1000 cycles	4–6 years	19.5 mΩ	11.25 mΩ
592–054	3 × D Stick	(full spec)				19.5 mΩ	11.25 mΩ
Ni-Cad Mass Plate		(see note 4)					(see note 2)
591–477	PCB Battery	Max limits −20°C−+50°C	180 mA 180 mA			1.5 Ω 3.5 Ω	930 mΩ 2.17 Ω
591–089	PP3		300 mA	300 cycles	5 years	375 mΩ	190 mΩ
591–168	Button Cell	0°C−+45°C recommended	410 mA 200 mA			200 mΩ 1.87 Ω	100 mΩ 850 mΩ
591–174	Button Cell		410 mA			1.4 Ω	700 mΩ
591–180	Stack						
591–196	Stack						

Notes:
1. Safe maximum discharge currents.
2. At 1 kHz.
3. A maximum of 75°C is permissible for up to 24 hours.
4. At temperatures below 0°C maximum discharge is C/2.
5. Continuous/pulsed recommended maximum discharge currents.

Appendix E: Small signal diodes

Signal diodes

Axial lead
a) L = 4.25, Dia = 1.85
b) L = 5.2, Dia = 2.7
d) L = 3.81, Dia = 1.71
e) L = 7.6, Dia = 2.7
f) L = 2.6, Dia = 1.7

Coloured band indicates cathode

(k) TO-92
TO-18
1 2 3
g) k — a
h) a case k
j) a — k

Application Code:
General Purpose — Switching — ● High speed
VHF Tuner — Low leakage/low capacitance

Package	Order Code	V_{RRM} max (V)	$I_{F(AV)}$ max (mA)	I_F max (mA)	V_F max (V)	App'n Code	Order Code	V_{RRM} max (V)	$I_{F(AV)}$ max (mA)	I_F max (mA) @	V_F max (V)	App'n Code	Order Code
(e)	AAZ15 ■	100	140	250	0.8								
(e)	AAZ17 ■	75	140	250	0.8								
(a)	BA317	30	100	100	1.1		(a)	75	100	1.0			BAW62
(a)	BA482	30	100	200	1.2		(a)	50	75	1.0			BAX1301
(f)	BAS45	35	225	1	1.0		(a)	150	200	1.3			BAX16
(f)	BAT41 ◆	125	100	10	0.45		(a)	150	200	1.3			BAX1601
(a)	BAT42 ◆	30	100	15	0.4		(a)	150	200	1.3			BAX16ES
(a)	BAT43 ◆	30	100	10	0.45		(a)	100	200	1.0			BAY72
(a)	BAT46 ◆	100	150		0.45		(b)	100	200	0.8			BAY73 ◆
(a)	BAT47 ◆	20	350		0.4		(b)	20	1000	0.55	1000		BYV10-20 ◆
							(b)	30	1000	0.55	1000		BYV10-30 ◆
(a)	BAT48 ◆	40	350	10	0.4	●	(b)	40	1000	0.55	1000		BYV10-40 ◆
(b)	BAT49 ◆	80	1000	100	0.42	●	(b)	60	1000	0.7	1000		BYV10-60 ◆
(f)	BAT81 ◆	40	30	1	0.41	●	(k)	35	10	1.5	5		JPAD100
(f)	BAT83 ◆	60	30	1	0.41		(k)	35	10	1.5	5		JPAD50
(f)	BAT85 ◆	30	200	10	0.4	●	(j)	35	50	1.5	5		PAD100
							(h)	45	50	1.5	5		PAD5
(a)	BAV10	60	300	200	1.0								
(a)	BAV19	120	250	200	1.2		(e)	30	110	0.54	130		OA47 ■
(a)	BAV20	200	250	200	1.2		(e)	30	10	2.0	30		OA9005 ■
(a)	BAV21	250	250	200	1.2		(e)	115	50	2.1	30		OA9105 ■
(g)	BAV45	35	50	10	1.0		(e)	115	50	1.85	30		OA9505 ■
							(e)	150	80	1.15	30		OA20201 ■
							(e)	400VRW	400	1.0	400		ZS104
							(a)	100	75	1.0	10		1N914
							(a)	125	200	0.8	10		1N916
							(a)	75	150	1.0	10		1N3595
							(a)	75	200	1.0	10		1N4148
							(a)	75	75	1.0	10		1N4148-NSC
							(a)	50	75	0.74	10		1N4149
							(d)	75	150	1.0	10		1N4150
							(d)	75	75	1.0	20		1N4446
							(d)	75	75	1.0	100		1N4448
							(d)	40	75	1.0	100		1N4448-NSC
							(d)	50	200	1.0	10		1S44
							(d)	100	200	1.2	200		1S920
							(d)	150	200	1.2	200		1S921
													1S922
							(d)	200	200	1.2	200		1S923

■ Germanium ◆ Silicon schottky barrier

Appendix F: Power diodes

Power diodes

Package (Not relative size)	IF (AV) Max Mean F'ward Current	50-100	200	300	400	600	800	1000	1200	1600
L=4.6 D=3.8 GLASS	1A	1N4001 (50 V)			1N5060					
L=5 D=2.7 PLASTIC	1A	1N4002 (100 V)	1N4003		1N4004	1N4005	1N4006	1N4007		
		1N4001TR■ (50 V)	1N4003TR■		1N4004TR■	1N4005TR■	1N4006TR■	1N4007TR■		
		1N4002TR■ (100 V)									
		1N4001GP◆ (50 V)	1N4003GP◆		1N4004GP◆	1N4005GP◆	1N4006GP◆	1N4007GP◆		
		1N4002GP◆ (100 V)									
L=6.35 D=6.35 GLASS	3A		1N5624			1N5626				
L=8.9 D=3.7 PLASTIC	3A	30S1(100 V)	30S2		30S4	30S6	30S8	30S10		
L=9.65 D=5.3 PLASTIC	3A	1N5401(100 V)	1N5402		1N5404	1N5406		1N5408		
			MR502		MR504					
L=9.1 D=9.1 PLASTIC	6A	G1750(50 V)	G1752			G1756				
		G1751(100 V)								
L=9.5 D=6.35 PLASTIC	6A	60S1D10DE (100 V)	60S2		60S4	60S6	60S8	60S10		
TO-220	6.5A			BY249-300		BY249-600				

Power diodes continued

Current		100V	200V	300V	400V	600V	800V	1000V	1200V	1600V
10A				BYX98–300**		BYX98–600**			BYX98–1200**	
				BYX98–300R*		BYX98–600R*			BYX98–1200R*	
12A			12F20**					12F100**		
			12FR20*							
15A				BYX99–300**					BYX99–1200**	
16A		M16–100**	M16–200**		M16–400**		M16–800**		M16–1200**	
		M16–100R*	M16–200R*		M16–400R*		M16–800R*		M16–1200R*	
			16F20**				16F80**	16F100**	16F120**	
								16FR100*	16FR120*	
30A				BYX96–300**		BYX96–600**			BYX96–1200**	BYX96–1600**
				BYX96–300R*		BYX96–600R*			BYX96–1200R*	BYX96–1600R*
40A	(A)	M41–100**	M41–200**			M41–600**				
	(A)	M41–100R*	M41–200R*			M41–600R*				
	(A)	40HF10**	40HF20**		40HF40**	40HF60**	40HF80**	40HF100**	40HF120**	
	(A)	40HFR10*	40HFR20*		40HFR40*	40HFR60*	40HFR80*	40HFR100*	40HFR120*	
47A	(B)					BYX97–600**			BYX97–1200**	BYX97–1600**
	(B)					BYX97–600R*			BYX97–1200R*	BYX97–1600R*
70A	(A)	M71–100**	M71–200**		M71–400**	M71–600**	M71–800**			
	(A)	M71–100R*	M71–200R*		M71–400R*	M71–600R*	M71–800R*			
	(A)	70HF10**	70HF20**		70HF40**	70HF60**	70HF80**	70HF100**	70HF120**	
	(A)	70HFR10*	70HFR20*		70HFR40*	70HFR60*	70HFR80*	70HFR100*	70HFR120*	
150A	(A)	45L10**	45L20**		45L40**	45L60**	45L80**	45L100**	45L120**	
	(A)	45LR10*	45LR20*		45LR40*	45LR60*			45LR120*	
250A	(B)	70U10**	70U20**		70U40**				70U120**	
	(B)	70UR10*	70UR20*						70UR120*	

10-32 UNF 2A

METRIC M5

(A) ¼-28 UNF 2A
(B) METRIC M6

(A) ½-20 UNF 2A (B) ⅝-16 UNF 2A

****Denotes stud cathode
*Denotes stud anode
■ Denotes bandoliered

◆ Glass passivated hermetically sealed construction with proven reliability equal to MIL-S-19500

Important — Forward current ratings quoted on stud mounting devices are maximum rating. Manufacturer's data should always be consulted as in some cases devices have to be forced air cooled to obtain the maximum ratings quoted

Appendix G: Zener Diodes

Zener diodes

L = 4.5, D = 2.0 BZX55 Series — glass DO35
L = 4.25, D = 1.85 BZY79 Series — glass DO35

L = 12.5, D = 6.5
BZX79 Series — plastic
Rounded end indicates cathode

L = 4.8, D = 2.6 BZV85 Series — glass DO41
L = 5.2, D = 2.7 BZX85 Series — glass DO41

Axial lead types: Coloured band indicates cathode

L = 8.9, D = 3.7
1N5000 Series — plastic

BZY93 Series — 10/32 UNF 2A stud

L = 4.57, D = 3.81 BZT03 Series — glass SOD-57
L = 5.0, D = 4.5 BZW03 Series — glass SOD-64

BZY91 Series ¼ × 28 UNF Stud

These types are available as normal (stud cathode). Add suffix 'R' to Order Code if reverse polarity is required.

Mftr	Philips	SGS-Thomson	SGS-Thomson	Philips	Philips	Philips	—	Philips	Philips	Philips
Nominal Zener Voltage	400mW	500mW	1.3W	1.3W	2.5W	3W	5W	6W	20W	75W
2.4 V	BZX79C2V4	BZX55C2V4	BZX85C2V4							
2.7 V	BZX79C2V7	BZX55C2V7	BZX85C2V7							
3 V	BZX79C3V0	BZX55C3V0	BZX85C3V0							
3.3 V	BZX79C3V3	BZX55C3V3	BZX85C3V3							
3.6 V	BZX79C3V6	BZX55C3V6	BZX85C3V6				1N5333B			
3.9 V	BZX79C3V9	BZX55C3V9	BZX85C3V9				1N5334B			
4.3 V	BZX79C4V3	BZX55C4V3	BZX85C4V3				1N5335B			
4.7 V	BZX79C4V7	BZX55C4V7	BZX85C4V7				1N5336B			
5.1 V	BZX79C5V1	BZX55C5V1	BZX85C5V1	BZV85C5V1			1N5337B			
5.6 V	BZX79C5V6	BZX55C5V6	BZX85C5V6	BZV85C5V6			1N5338B			
6.2 V	BZX79C6V2	BZX55C6V2	BZX85C6V2	BZV85C6V2			1N5339B			
6.8 V	BZX79C6V8	BZX55C6V8	BZX85C6V8	BZV85C6V8			1N5341B			
7.5 V	BZX79C7V5	BZX55C7V5	BZX85C7V5	BZV85C7V5	BZX70C7V5	BZT03C7V5	1N5342B		BZY93C7V5#	
8.2 V	BZX79C8V2	BZX55C8V2	BZX85C8V2	BZV85C8V2	BZX70C8V2	BZT03C8V2	1N5343B		BZY93C8V2#	
9.1 V	BZX79C9V1	BZX55C9V1	BZX85C9V1	BZV85C9V1	BZX70C9V1	BZT03C9V1	1N5344B		BZY93C9V1#	
10 V	BZX79C10	BZX55C10	BZX85C10	BZV85C10	BZX70C10	BZT03C10	1N5346B	BZW03-C10 NEW	BZY93C10#	BZY91C7V5
11 V	BZX79C11	BZX55C11	BZX85C11	BZV85C11	BZX70C11	BZT03C11	1N5347B		BZY93C11	
12 V	BZX79C12	BZX55C12	BZX85C12	BZV85C12	BZX70C12	BZT03C12	1N5348B	BZW03-C12 NEW	BZY93C12#	BZY91C10
13 V	BZX79C13	BZX55C13	BZX85C13	BZV85C13	BZX70C13	BZT03C13	1N5349B		BZY93C13#	BZY91C12
15 V	BZX79C15	BZX55C15	BZX85C15	BZV85C15	BZX70C15	BZT03C15	1N5350B		BZY93C15#	BZY91C15#

WATTAGE (All ± 5% Voltage Tolerance)

Zener diodes continued

16 V	BZX79C16	BZX55C16	BZX85C16	BZV85C16	BZX70C16	BZT03C16			BZY93C16#	
18 V	BZX79C18	BZX55C18	BZX85C18	BZV85C18	BZX70C18	BZT03C18	1N5353B		BZY93C18#	BZY91C18#
20 V	BZX79C20	BZX55C20	BZX85C20	BZV85C20	BZX70C20	BZT03C20	1N5355B		BZY93C20#	BZY91C20
22 V	BZX79C22	BZX55C22	BZX85C22	BZV85C22	BZX70C22	BZT03C22	1N5357B		BZY93C22	
24 V	BZX79C24	BZX55C24	BZX85C24	BZV85C24	BZX70C24	BZT03C24	1N5358B	BZW03-C24 NEW	BZY93C24#	BZY91C24
27 V	BZX79C27	BZX55C27	BZX85C27	BZV85C27	BZX70C27	BZT03C27	1N5359B	BZW03-C27 NEW	BZY93C27#	BZY91C27
30 V	BZX79C30	BZX55C30	BZX85C30	BZV85C30	BZX70C30	BZT03C30	1N5361B		BZY93C30#	BZY91C30
33 V	BZX79C33	BZX55C33	BZX85C33	BZV85C33	BZX70C33	BZT03C33	1N5363B		BZY93C33#	BZY91C33
36 V	BZX79C36	BZX55C36	BZX85C36	BZV85C36	BZX70C36	BZT03C36	1N5364B	BZW03-C36 NEW	BZY93C36	BZY91C36
39 V	BZX79C39	BZX55C39	BZX85C39	BZV85C39	BZX70C39	BZT03C39	1N5365B		BZY93C39#	
43 V	BZX79C43	BZX55C43	BZX85C43	BZV85C43	BZX70C43	BZT03C43	1N5366B		BZY93C43	BZY91C43
47 V	BZX79C47	BZX55C47	BZX85C47	BZV85C47	BZX70C47	BZT03C47	1N5367B	BZW03-C47 NEW	BZY93C47	BZY91C47
51 V	BZX79C51	BZX55C51	BZX85C51	BZV85C51	BZX70C51	BZT03C51	1N5368B	BZW03-C51 NEW	BZY93C51	BZY91C51
56 V	BZX79C56	BZX55C56	BZX85C56	BZV85C56	BZX70C56	BZT03C56	1N5369B		BZY93C56	
62 V	BZX79C62	BZX55C62	BZX85C62	BZV85C62	BZX70C62	BZT03C62	1N5370B		BZY93C62	BZY91C62
68 V	BZX79C68	BZX55C68	BZX85C68	BZV85C68	BZX70C68	BZT03C68	1N5372B		BZY93C68	BZY91C68
75 V	BZX79C75	BZX55C75	BZX85C75	BZV85C75	BZX70C75	BZT03C75	1N5373B	BZW03-C75 NEW	BZY93C75#	BZY91C75
82 V		BZX55C82	BZX85C82			BZT03C82	1N5374B	BZW03-C82 NEW		
91 V		BZX55C91	BZX85C91			BZT03C91	1N5375B			
100 V		BZX55C100	BZX85C100			BZT03C100	1N5377B			
110 V		BZX55C110	BZX85C110			BZT03C110	1N5378B			
120 V		BZX55C120	BZX85C120			BZT03C120	1N5379B			
130 V		BZX55C130	BZX85C130			BZT03C130	1N5380B			
150 V		BZX55C150	BZX85C150			BZT03C150	1N5381B			
160 V		BZX55C160	BZX85C160			BZT03C160	1N5383B			
180 V		BZX55C180	BZX85C180			BZT03C180	1N5384B			
200 V		BZX55C200	BZX85C200			BZT03C200	1N5386B			
220 V						BZT03C220	1N5388B			
240 V						BZT03C240				
270 V						BZT03C270				

EPOXY-POTTED BRIDGE RECTIFIERS

Voltage	Current	Device No.
200	2	KBPC 102
400	2	KBPC 104
600	2	KBPC 106
800	2	KBPC 108
200	4	KBU 4D
800	4	KBU 4K
200	6	KBPC 802
800	6	KBPC 808
200	12	SKB 25/02
800	12	SKB 25/08
1200	12	SKB 25/12
50	25	KBPC 25005
200	25	KBPC 2502
600	25	KBPC 2506
200	35	KBPC 3502
600	35	KBPC 3506

Notes:
1. The bridge assembly should be mounted on a heat sink.
2. Current ratings are for resistive loads. When the rectifier is used on a battery or capacitive load the current rating should be multiplied by 0.8.

Appendix H: Transistors

Transistor pin connections

TO18, TO5, TO39, TO205

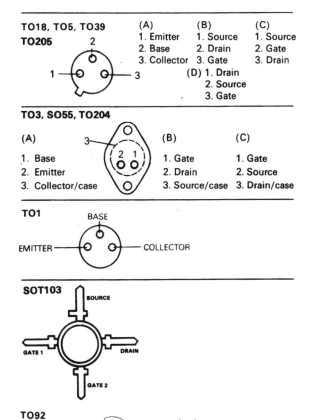

(A)
1. Emitter
2. Base
3. Collector

(B)
1. Source
2. Drain
3. Gate

(C)
1. Source
2. Gate
3. Drain

(D)
1. Drain
2. Source
3. Gate

TO3, SO55, TO204

(A)
1. Base
2. Emitter
3. Collector/case

(B)
1. Gate
2. Drain
3. Source/case

(C)
1. Gate
2. Source
3. Drain/case

TO1

BASE, EMITTER, COLLECTOR

SOT103

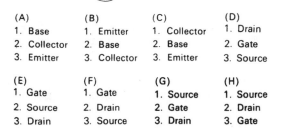

SOURCE, GATE 1, DRAIN, GATE 2

TO92, TO237

plastic

(A)
1. Base
2. Collector
3. Emitter

(B)
1. Emitter
2. Base
3. Collector

(C)
1. Collector
2. Base
3. Emitter

(D)
1. Drain
2. Gate
3. Source

(E)
1. Gate
2. Source
3. Drain

(F)
1. Gate
2. Drain
3. Source

(G)
1. Source
2. Gate
3. Drain

(H)
1. Source
2. Drain
3. Gate

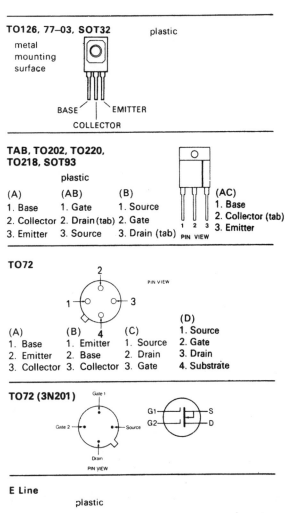

TO126, 77-03, SOT32
plastic

metal mounting surface

BASE, EMITTER, COLLECTOR

TAB, TO202, TO220, TO218, SOT93
plastic

(A)
1. Base
2. Collector
3. Emitter

(AB)
1. Gate
2. Drain (tab)
3. Source

(B)
1. Source
2. Gate
3. Drain (tab)

(AC)
1. Base
2. Collector (tab)
3. Emitter

PIN VIEW

TO72

PIN VIEW

(A)
1. Base
2. Emitter
3. Collector

(B)
1. Emitter
2. Base
3. Collector

(C)
1. Source
2. Drain
3. Gate

(D)
1. Source
2. Gate
3. Drain
4. Substrate

TO72 (3N201)

Gate 1, Gate 2, Source, Drain

PIN VIEW

E Line
plastic

COLLECTOR
BASE
EMITTER

type	material	case	application	P_T	I_C	V_{CEO}	V_{CBO}	h_{FE}	$I_T(typ)$
AC127	NPN Ge	TO1	Audio output	340 mW	500 mA	12 V	32 V	50	2.5 MHz
AC128	PNP Ge	TO1	Audio output	700 mW	−1 A	−16 V	−32 V	60−175	1.5 MHz
AD149	PNP Ge	TO3(A)	Audio output	*22.5 W at 50°C	−3.5 A	−50 V	−50 V	30−100	0.5 MHz
AD161 } Pair	NPN Ge }	SO55(A)	Audio matched pair	*4 W at 72°C *6 W at 63°C	3 A	20 V	32 V	50−300	3 MHz
AD162 }	PNP Ge }				−3 A	−20 V	−32 V	50−300	1.5 MHz
AF127	PNP Ge	TO72(A)	I.F. Applications	60 mW	−1.0 mA	−20 V	−20 V	150	75 MHz
BC107	NPN Si	TO18	Audio driver stages (complement BC177)	360 mW	100 mA	45 V	50 V	110−450	250 MHz
BC108	NPN Si	TO18	General purpose (complement BC178)	360 mW	100 mA	20 V	30 V	110−800	250 MHz
BC109	NPN Si	TO18	Low noise audio (complement BC179)	360 mW	100 mA	20 V	30 V	200−800	250 MHz
BC142	NPN Si	TO39	Audio driver	800 mW	800 mA	60 V	80 V	20 (min.)	80 MHz
BC143	PNP Si	TO39	Audio driver	800 mW	−800 mA	−60 V	−60 V	25 (min.)	160 MHz
BC177	PNP Si	TO18	Audio driver stages (complement BC107)	300 mW	−100 mA	−45 V	−50 V	125−500	200 MHz
BC178	PNP Si	TO18	General purpose (complement BC108)	300 mW	−100 mA	−25 V	−30 V	125−500	200 MHz
BC179	PNP Si	TO18	Low Noise Audio (complement BC109)	300 mW	−100 mA	−20 V	−25 V	240−500	200 MHz
BC182L	NPN Si	TO92(A)	General purpose	300 mW	200 mA	50 V	60 V	100−480	150 MHz
BC183L	NPN Si	TO92(A)	General purpose (complement BC213L)	300 mW	200 mA	30 V	45 V	100−850	280 MHz
BC184L	NPN Si	TO92(A)	General purpose	300 mW	200 mA	30 V	45 V	250 (min.)	150 MHz
BC212L	PNP Si	TO92(A)	General purpose	300 mW	−200 mA	−50 V	−60 V	60−300	200 MHz
BC213L	PNP Si	TO92(A)	General purpose (complement BC183L)	300 mW	−200 mA	−30 V	−45 V	80−400	350 MHz
BC214L	PNP Si	TO92(A)	General purpose	300 mW	−200 mA	−30 V	−45 V	140−600	200 MHz
BC237B	NPN Si	TO92(B)	Amplifier	350 mW	100 mA	45 V	50 V	120−800	100 MHz
BC307B	PNP Si	TO92(B)	Amplifier	350 mW	100 mA	45 V	50 V	120−800	280 MHz
BC327	PNP Si	TO92(B)	Driver	625 mW	−500 mA	−45 V	−50 V	100−600	260 MHz
BC337	NPN Si	TO92(B)	Audio driver	625 mW at 45°C	500 mA	45 V	50 V	100−600	200 MHz
BC441	NPN Si	TO39	General purpose (complement BC461)	1 W	2 A peak	60 V	75 V	40−250	50 MHz (min.)
BC461	PNP Si	TO39	General purpose (complement BC441)	1 W	−2 A peak	−60 V	−75 V	40−250	50 MHz
BC477	PNP Si	TO18	Audio driver stages	360 mW	−150 mA	−80 V	−80 V	110−950	150 MHz
BC478	PNP Si	TO18	General purpose	360 mW	−150 mA	−40 V	−40 V	110−800	150 MHz
BC479	PNP Si	TO18	Low noise audio amp.	360 mW	−150 mA	−40 V	−40 V	110−800	150 MHz
BCY70	PNP Si	TO18	General purpose	360 mW	−200 mA	−40 V	−50 V	150	200 MHz
BCY71	PNP Si	TO18	Low noise general purpose	360 mW	−200 mA	−45 V	−45 V	100−400	200 MHz

BD131	NPN Si	TO126	General purpose—medium power	15 W at 60°C	3 A	45 V	70 V	20 (min.)	60 MHz
BD132	PNP Si	TO126	General purpose—medium power	15 W at 60°C	−3 A	−45 V	−45 V	20 (min.)	60 MHz
BD131 ⎫ Pair **BD132** ⎭	NPN Si ⎫ PNP Si ⎭	TO126	Audio matched pair	15 W at 60°C 15 W at 60°C	3 A −3 A	45 V −45 V	70 V −45 V	20 (min.) 20 (min.)	60 MHz 60 MHz
BD135	NPN Si	SOT32	Audio driver	12.5 W at 25°C	1.5 A	45 V	45 V	40–250	50 MHz
BD136	PNP Si	SOT32	Audio driver	12.5 W at 25°C	−1.5 A	−45 V	−45 V	40–250	75 MHz
BD437 **BD438**	NPN Si PNP Si	TO126	Power switching complementary	36 W at 25°C	4 A	45 V	45 V	40	3 MHz
BCY70	PNP Si	TO18	General purpose	360 mW	−200 mA	−40 V	−50 V	150	200 MHz
BCY71	PNP Si	TO18	Low noise general purpose	360 mW	−200 mA	−45 V	−45 V	100–400	200 MHz
BD131	NPN Si	TO126	General purpose—medium power	15 W at 60°C	3 A	45 V	70 V	20 (min.)	60 MHz
BD132	PNP Si	TO126	General purpose—medium power	15 W at 60°C	−3 A	−45 V	−45 V	20 (min.)	60 MHz
BD131 ⎫ Pair **BD132** ⎭	NPN Si ⎫ PNP Si ⎭	TO126	Audio matched pair	15 W at 60°C 15 W at 60°C	3 A −3 A	45 V −45 V	70 V −45 V	20 (min.) 20 (min.)	60 MHz 60 MHz
BD135	NPN Si	SOT32	Audio driver	12.5 W at 25°C	1.5 A	45 V	45 V	40–250	50 MHz
BD136	PNP Si	SOT32	Audio driver	12.5 W at 25°C	−1.5 A	−45 V	−45 V	40–250	75 MHz
BD437 **BD438**	NPN Si PNP Si	TO126	Power switching complementary	36 W at 25°C	4 A	45 V	45 V	40	3 MHz
BD679 **BD680**	NPN Si PNP Si	TO126	Audio complementary Darlington	40 W at 25°C 40 W at 25°C	6 A 6 A	80 V 80 V	100 V 80 V	2200 2200	60 kHz 60 kHz
BDX33C **BDX34C**	NPN Si PNP Si	TO220(A) (A)	Power switching Darlington	70 W at 25°C	10 A	100 V	100 V	750 (min.)	60 kHz
BF259	NPN Si	TO39	High-voltage video amplifier	800 mW	100 mA	300 V	300 V	25	90 MHz
BF337	NPN Si	TO39	Video amplifier	3 W at 125°C	100 mA	200 V	250 V	20 (min.)	80 MHz
BFY50	NPN Si	TO39	High voltage general purpose	800 mW	1 A	35 V	80 V	30	60 MHz
BFY51	NPN Si	TO39	General purpose	800 mW	1 A	30 V	60 V	40	50 MHz
BFY52	NPN Si	TO39	General purpose	800 mW	1 A	20 V	40 V	60 V	50 MHz

Type	Polarity	Package	Description	Power	Current	V(CEO)	V(CBO)	hFE	fT
TIP31A	NPN Si	TO220(A)	Plastic medium power complementary	40 W at 25°C	3 A	60 V	60 V	10–60	8 MHz
TIP32A	PNP Si							10–40	
TIP31C	NPN Si	TO220(A)	Plastic medium power complementary	40 W at 25°C	3 A	100 V	100 V	10–50	8 MHz
TIP32C	PNP Si								
TIP33A	NPN Si	TAB(A)	Audio output complementary	80 W at 25°C	10 A	60 V	60 V	20–100	3 MHz
TIP34A	PNP Si								
2N2905	PNP Si	TO5(A)	Switching	600 mW	−600 mA	−40 V	−60 V	100–300	200 MHz (min.)
2N2905A	PNP Si	TO5(A)	Switching	600 mW	−600 mA	−60 V	−60 V	100–300	200 MHz (min.)
2N2907A	PNP Si	TO18(A)	Switching	400 mW	−600 mA	−60 V	−60 V	100–300	200 MHz (min.)
2N3019	NPN Si	TO39(A)	General purpose	500 mW	1 A	80 V	140 V	90 min.	100 MHz
2N3053	NPN Si	TO39(A)	General purpose	800 mW	1 A	40 V	60 V	50–250	100 MHz
2N3055E	NPN Si	TO3(A)	High power epitaxial (complement MJ2955)	115 W at 25°C	15 A	60 V	100 V	20–70	2.5 MHz
2N3055H	NPN Si	TO3(A)	High power homotaxial (complement PNP3055)	115 W at 25°C	15 A	60 V	100 V	20–70	1 MHz
PNP3055	PNP Si	TO3(A)	High power (complement 2N3055H)	115 W at 25°C	−15 A	−60 V	−100 V	20–70	0.8 MHz
2N3440	NPN Si	TO39(A)	General purpose	1 W	1 A	250 V	300 V	40–160	15 MHz
2N3702	PNP Si	TO92(A)	General purpose	360 mW	−200 mA	−25 V	−40 V	60–300	100 MHz

Appendix I: Voltage regulators

Voltage regulators

Fixed voltage series regulators

Description	Output Voltage	Current	Case	Stock No.	Suitable transformer Stock No.	Sec. Voltage	VA
78L05 RS309K (LM309K)	5 V	100 mA	TO92	306-190	207-188	9(S)	6
7805	5 V	1.2 A	TO3	305-614	207-122	9(S)	20
78S05	5 V	1.0 A	TO220	305-888	207-122	9(S)	20
78H05	5 V	2 A	TO220	633-026	207-239	9(S)	50
78L12	5 V	5 A	TO3	307-301	207-239	9(S)	50
7812	12 V	100 mA	TO92	306-207	207-217	15(P)	6
78S12	12 V	1.0 A	TO220	305-894	207-267	15(P)	50
78H12	12 V	2 A	TO220	633-032	207-267	15(P)	50
78L15	12 V	5 A	TO3	307-317	207-289	12(S)	100
7815	15 V	100 mA	TO92	306-213	207-217	15(P)	6
78S15	15 V	1.0 A	TO220	305-901	207-267	15(P)	50
78L24	15 V	2 A	TO220	633-048	207-267	15(P)	50
7824	24 V	100 mA	TO92	306-229	207-201	24(S)	6
78S24	24 V	1.0 A	TO220	305-917	207-251	24(S)	50
79L05	24 V	2 A	TO220	633-054	207-295	24(S)	100
7905	-5 V	-100 mA	TO92	306-235	207-188	9(S)	6
79L12	-5 V	-1.2 A	TO220	306-049	207-122	9(S)	20
7912	-12 V	-100 mA	TO92	306-241	207-217	15(P)	6
79L15	-12 V	-1.2 A	TO220	306-055	207-267	15(P)	50
7915	-15 V	-100 mA	TO92	306-257	207-217	15(P)	6
79L24	-15 V	-1.2 A	TO220	305-923	207-267	15(P)	50
7924	24 V	-100 mA	TO92	306-263	207-201	24(S)	6
	-24 V	-1.0 A	TO220	306-184	207-251	24(S)	50

Source: RS Data Library

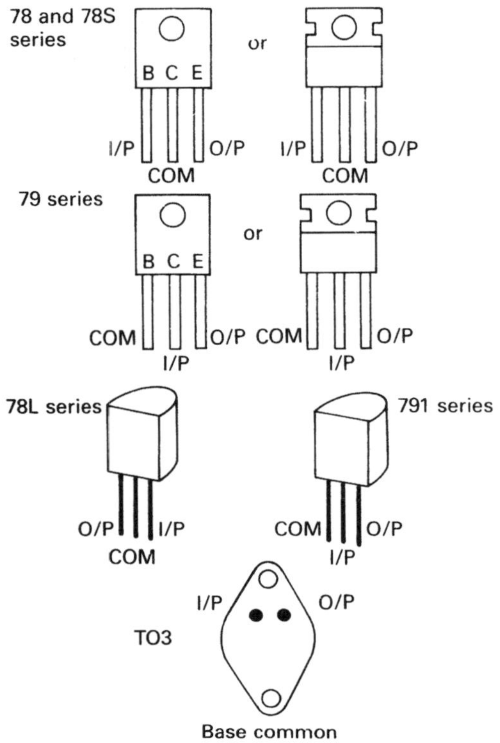

78H05 AND 78H12 FIXED HYBRID REGULATORS

Two fixed voltage hybrid regulators, housed in T03 style metal cases, capable of supplying output currents up to 5 A. The internal circuitry limits the junction temperature to a safe value and provides automatic thermal overload protection.

Safe operating operating protection is also incorporated making the regulators virtually damage proof.

In order to achieve maximum performance the internal power dissipation must be kept below 50 W. Transformer and heatsink selections are dependent upon the exact application.

78H – BASIC CIRCUIT, FIXED VOLTAGE

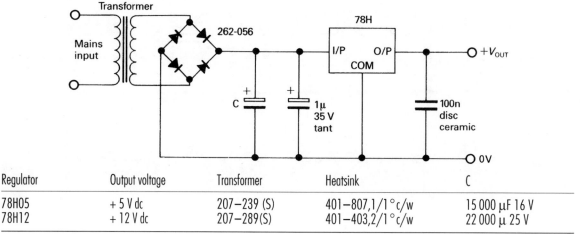

Regulator	Output voltage	Transformer	Heatsink	C
78H05	+5 V dc	207–239 (S)	401–807,1/1°c/w	15 000 µF 16 V
78H12	+12 V dc	207–289 (S)	401–403,2/1°c/w	22 000 µ 25 V

Source: RS Data Library

79 Series negative regulators

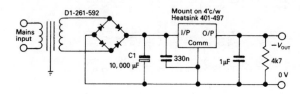

78 Series positive regulators

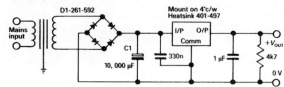

Constant current generator

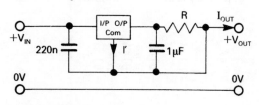

Source: RS Data Library

For T03 types $I' = 10$ mA typ.
For 78S series $I' = 8$ mA typ.
For 78/79 series $I' = 4.5$ mA typ.
For 78L/79L series $I' = 3.5$ mA typ.

Circuit gives constant current through load provided V_{OUT} does not exceed $V_{IN} - (V_R + 2.5)$. [$V_{IN} - (V_R + 3)$ for 78S series.] Select R to give designed constant current I_{OUT}

$$I_{OUT} = \frac{V_R}{R} + I'$$

where V_R is the basic regulator voltage.

INCREASING BASIC REGULATOR VOLTAGE

The input voltage V_{IN} should be derived from a suitable transformer, rectifier and smoothing capacitor circuit. Note V_{IN} must be greater (within maximum ratings) than $V_{OUT} + 2.5$ V.

Figure A1 gives higher output voltage than basic circuit but with reduced regulation.

$$V_{OUT} = V_R \left(1 + \frac{R_2}{R_1}\right) + I' R_2$$

$$I_{R1} \geq 5 \times I'$$

where V_R = basic regulator voltage
I' = 10 mA (TO3)
= 8 mA (78S series)
= 4.5 mA (78/79 series)
= 3.5 mA (78L/79L series)

Figure A2 gives better regulation than Fig. A1.

$$V_{OUT} = V_R + V_1 + 0.6$$

where $V_1 = \dfrac{R_2 \, V_{OUT}}{R_1 + R_2}$

and $\dfrac{R_1}{R_2} = \dfrac{V_R + 0.6}{V_{OUT} - (V_R + 0.6)}$

e.g. For 9 V output with 5 V regulator

$$R_1 = 5\,k6 \quad R_2 = 3\,k3$$

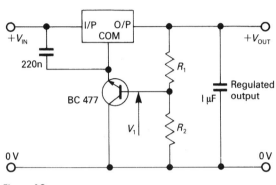

Figure A2

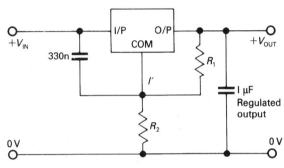

Figure A1

Appendix J: Power control

Power control, thyristor ratings

Device No.	Max. volts	Max. current	Gate I	Gate V
C203YY	60 V	0.8 A	0.2 mA	0.8 V
BTX18-400	500 V	1.0 A	5 mA	2 V
BT106	700 V	1.0 A	50 mA	3.5 V
C106	400 V	2.55 A	0.2 mA	0.8 V
2N4443	400 V	5.1 A	30 mA	1.5 V
2N4444	600 V	5.1 A	30 mA	1.5 V
BT152-600	600 V	13 A	32 mA	1 V
BTY79-400R	400 V	6.4 A	30 mA	3 V
BTY79-800R	800 V	6.4 A	30 mA	3 V
N018RH05	500 V	21 A	100 mA	3 V
N018RH08	800 V	21 A	100 mA	3 V
N018RH12	1200 V	21 A	100 mA	3 V
N029RH05	500 V	30 A	100 mA	3 V
N029RH08	800 V	30 A	100 mA	3 V
N029RH12	1200 V	30 A	100 mA	3 V
N044RH05	500 V	45 A	100 mA	3 V
N044RH08	800 V	45 A	100 mA	3 V
N044RH12	1200 V	45 A	100 mA	3 V
N060RH06	600 V	63 A	100 mA	3 V
N060RH08	800 V	63 A	100 mA	3 V
N060RH12	1200 V	63 A	100 mA	3 V
N086RH06	600 V	85 A	150 mA	3 V
N086RH08	800 V	85 A	150 mA	3 V
N086RH12	1200 V	85 A	150 mA	3 V
N105RH06	600 V	110 A	150 mA	3 V
N105RH08	800 V	110 A	150 mA	3 V
N105RH12	1200 V	110 A	150 mA	3 V

Power control, triac ratings

Device No.	Case type	Max. volts	Max. current	Gate V
Z0105DA	T092	400 V	0.35 A	2.0 V
T1CP206D	T092	400 V	1.5 A	2.5 V
T1CP206M	T092	600 V	1.5 A	2.5 V
T1C206M	T0220AB	600 V	4.0 A	2.0 V
T1C216M	T0220AB	600 V	6.0 A	3.0 V
T1C225M	T0220AB	600 V	8.0 A	2.0 V
T1C226M	T0220AB	600 V	8.0 A	2.0 V
T1C236M	T0220AB	600 V	12.0 A	2.0 V
BT139	T0220AB	600 V	15.0 A	1.5 V
T1C246M	T0220AB	600 V	16.0 A	2.0 V

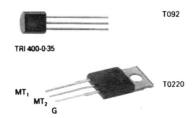

Appearance of a triac.

Diac trigger diodes

Device No.	Trigger volts	Max. current
BR100D0-14	32 V ± 4 V	2
133D0-7	32 V ± 4 V	1
D0201YR	32 V ± 4 V	1

Silicon bidirectional trigger diodes for use in triac firing circuits are glass encapsulated with the appearance of an axial lead small signal diode. (See Appendix E for photograph.)

Appendix K: Comparison of British and American logic gate symbols

Comparison of British and American logic gate symbols

Logic gate	American symbol	British symbol	Truth table
AND	A, B inputs → AND shape → Output	A, B → [&] → Output	A B OUT 0 0 0 0 1 0 1 0 0 1 1 1
OR	A, B inputs → OR shape → Output	A, B → [≥1] → Output	A B OUT 0 0 0 0 1 1 1 0 1 1 1 1
Exclusive - OR	A, B inputs → XOR shape → Output	A, B → [=1] → Output	A B OUT 0 0 0 0 1 1 1 0 1 1 1 0
NOT	A → triangle → Ā	A → [1] → Ā	A OUT 0 1 1 0
NOR	A, B → NOR shape → Output	A, B → [≥1]○ → Output	A B OUT 0 0 1 0 1 0 1 0 0 1 1 0
NAND	A, B → NAND shape → Output	A, B → [&]○ → Output	A B OUT 0 0 1 0 1 1 1 0 1 1 1 0

Appendix L: Integrated circuit logic gates

CMOS INTEGRATED CIRCUIT LOGIC GATES

4001B Quad 2 input NOR

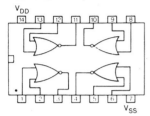

4002B Dual 4 input NOR

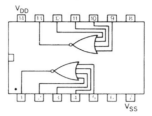

4023B Triple 3 input NAND

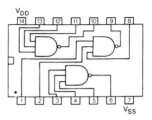

4011B Quad 2 input NAND

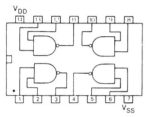

4012B Dual 4 input NAND

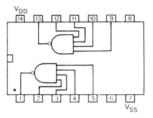

4013B Dual D type flip flop

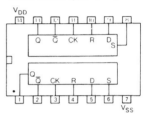

4025B Triple 3 input NOR

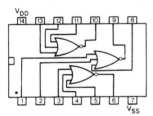

4027B Dual J.K. flip flop

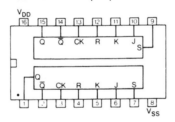

4049UB Hex inverter buffer

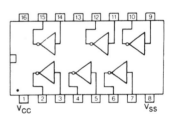

4069UB Hex inverter

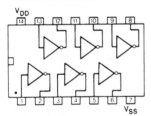

4071B Quad 2 input OR

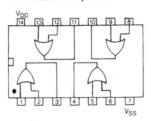

4072B Dual 4 input OR gate

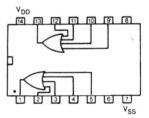

4073B Triple 3 input AND

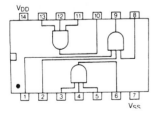

4075B Triple 3 input OR

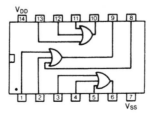

4081B Quad 2 input AND

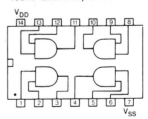

TTL INTEGRATED CIRCUIT LOGIC GATES

TTL integrated circuit logic gates

7400 Quadruple 2-input NAND gate
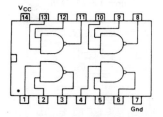

7402 Quadruple 2-input NOR gate
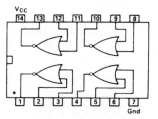

7404 Hex inverter
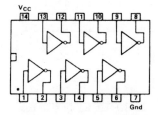

7408 Quadruple 2-input AND gate
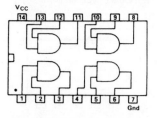

7410 Triple 3-input NAND gate
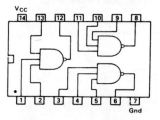

7411 Triple 3-input AND gate
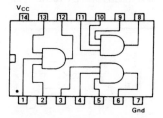

7414 Hex Schmitt Trigger
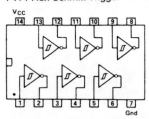

7420 Dual 4-input NAND gate
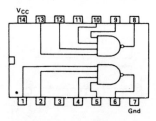

7421 Dual 4-input AND gate
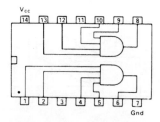

7427 Triple 3-input NOR gate
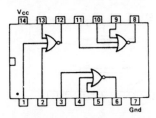

7432 Quadruple 2-input OR gate
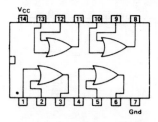

7474 Dual D-type edge-triggered Flip Flop
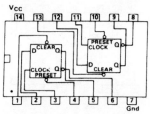

7476 Dual JK Flip Flop with set and clear
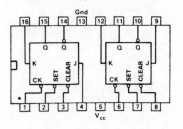

Appendix M: Thermocouple colour coding

Thermocouple colour coding

Generic and trade names	Colour Coding			Magnetic		Maximum useful temp. range	EMF (mV) over useful temp. range	Average sensitivity µV/°C	Environment (bare wire)
	Single	Overall T/C wire	Overall extension grade wire	Yes	No				
Copper Constantan, Cupron, Advance	Blue Red	Brown	Blue		X X	°F −328 to 662 °C −200 to 350	−5.602 to 17.816	40.5	Mild oxidising, reducing. Vacuum or inert. Good where moisture is present
Iron Constantan, Cupron, Advance	White Red	Brown	Black	X	X	°F 32 to 1382 °C 0 to 750	0 to 42.283	52.6	Reducing. Vacuum, inert. Limited use in oxidising at high temperatures. Not recommended for low temps
Chromel, Tophel, T¹ Thermokanthal KP Constantan, Cupron, Advance	Purple Red	Brown	Purple		X X	°F −328 to 1652 °C −200 to 900	−8.824 to 68.783	67.9	Oxidising or inert. Limited use in vacuum or reducing
Chromel, Tophel, T¹ Thermokanthal KP Alumel, Nial, T² Thermokanthal KN	Yellow Red	Brown	Yellow	X	X	°F −328 to 2282 °C −200 to 1250	−5.973 to 50.633	38.8	Clean oxidising and inert. Limited use in vacuum or reducing
Platinum 10% Rhodium Pure Platinum	Black Red		Green		X X	°F32 to 2642 °C 0 to 1450	0 to 14.973	10.6	Oxidising or inert atmos. Do not insert in metal tubes. Beware of contamination
Platinum 13% Rhodium Pure Platinum	Black Red		Green		X X	°F32 to 2642 °C 0 to 1450	0 to 16.741	12.0	
Platinum 30% Rhodium Platinum 6% Rhodium	Grey Red		Grey		X X	°F 32 to 3092 °C 0 to 1700	0 to 12.426	7.6	
Tungsten 5% Rhenium Tungsten 26% Rhenium	White/red trace Red		White/red trace		X X	°F 32 to 4208 °C 0 to 2320	0 to 37.066	16.6	Vacuum, inert, hydrogen atmospheres. Beware of embrittlement
Tungsten Tungsten 26% Rhenium	White/blue trace Red		White/blue trace		X X	°F 32 to 4208 °C 0 to 2320	0 to 38.414	16.0	
Tungsten 3% Rhenium Tungsten 25% Rhenium	White/yellow trace Red		White/yellow trace		X X	°F 32 to 4208 °C 0 to 2320	0 to 39.506	17.0	

Notes:
1. Standard ANSI colour coding is used on all insulated thermocouples wire and extension grade wire when the insulation permits. For some insulations a coloured tracer is used to distinguish the calibration.
2. Thermocouple and extension grade wires are specified by ANSI letter designations. Positive and negative legs are identified by the letter suffixes P and N respectively. Example: JP designates the positive leg (iron) of the Iron-Constantan pair.

Appendix N: Strain Gauges

Strain gauges

General specification (all types)
Measurable strain	2 to 4% maximum
Thermal output 20 to 160°C	± 2 micro strain/°C*
160 to 180°C	± 5 micro strain/°C*
Gauge factor change with temperature	±0.015%/°C max
Gauge resistance	120 Ω
Gauge resistance tolerance	±0.5%
Fatigue life	>10^5 reversals @ 1000 micro strain*
Foil material	copper nickel alloy

*1 micro strain is equivalent to an extension of 0.0001%

Specification (standard polyester backed types)
Temperature range	−30°C to +80°C
Gauge length	8 mm
Gauge width	2 mm
Gauge factor	2.1
Base length (single types)	13.0 mm
Base width (single types)	4.0 mm
Base diameter (rosettes)	21.0 mm

Specification (miniature polyimide backed type)
Temperature range	−30°C to +180°C
Gauge length	2 mm to 5 mm
Gauge width	1.6 mm to 1.8 mm
Gauge factor	2.0 to 2.1
Base length (single types)	6.0 mm to 9.0 mm
Base width (single types)	2.5 mm to 3.5 mm
Base size (rosettes)	7.5 × 7.5 mm, 12 × 12 mm

Appendix O: Health and Safety Executive (HSE) Publications and Information

HSE Books, Information Leaflets and Guides may be obtained from
HSE Books, P.O. Box 1999, Sudbury, Suffolk CO10 6FS

HSE Infoline – Telephone No. 01541 545500 or write to
HSE Information Centre, Broad Lane, Sheffield S3 7HO

HSE home page on the World Wide Web
http:/www.open.gov.uk/hse/hsehome.htm

Environmental Health Department of the Local Authority
Look in the local telephone directory under the name of the authority.

HSE AREA OFFICES

01 South West
Inter City House, Mitchell Lane, Victoria Street, Bristol BS1 6AN
Telephone: 01171 290681

02 South
Priestley House, Priestley Road,
Basingstoke RG24 9NW Telephone: 01256 473181

03 South East
3 East Grinstead House, London Road,
East Grinstead, West Sussex RH19 1RR
Telephone: 01342 326922

05 London North
Maritime House, 1 Linton Road, Barking,
Essex IG11 8HF Telephone: 0208 594 5522

06 London South
1 Long Lane London SE1 4PG
Telephone: 0207 407 8911

07 East Anglia
39 Baddow Road, Chelmsford, Essex CM2 OHL
Telephone: 01245 284661

08 Northern Home Counties
14 Cardiff Road, Luton, Beds LU1 1PP
Telephone: 01582 34121

09 East Midlands
Belgrave House, 1 Greyfriars, Northampton NN1 2BS
Telephone: 01604 21233

10 West Midlands
McLaren Building, 2 Masshouse Circus, Queensway
Birmingham B4 7NP
Telephone: 0121 200 2299

11 Wales
Brunel House, Nizalan Road, Cardiff CF2 1SH
Telephone: 02920 473777

12 Marches
The Marches House, Midway, Newcastle-under-Lyme,
Staffs ST5 1DT Telephone: 01782 717181

13 North Midlands
Birkbeck House, Trinity Square, Nottingham NG1 4AU
Telephone: 0115 470712

14 South Yorkshire
Sovereign House, 40 Silver Street, Sheffield S1 2ES
Telephone: 0114 739081

15 West and North Yorkshire
8 St Paul's Street, Leeds LS1 2LE
Telephone: 0113 446191

16 Greater Manchester
Quay House, Quay Street, Manchester M3 3JB
Telephone: 0161 831 7111

17 Merseyside
The Triad, Stanley Road, Bootle L20 3PG
Telephone: 01229 922 7211

18 North West
Victoria House, Ormskirk Road, Preston PR1 1HH
Telephone: 01772 59321

19 North East
Arden House, Regent Centre, Gosforth, Newcastle upon Tyne NE3 3JN
Telephone: 0191 284 8448

20 Scotland East
Belford House, 59 Belford Road, Edinburgh EH4 3UE
Telephone: 0181 225 1313

21 Scotland West
314 St Vincent Street, Glasgow G3 8XG
Telephone: 0141 204 2646

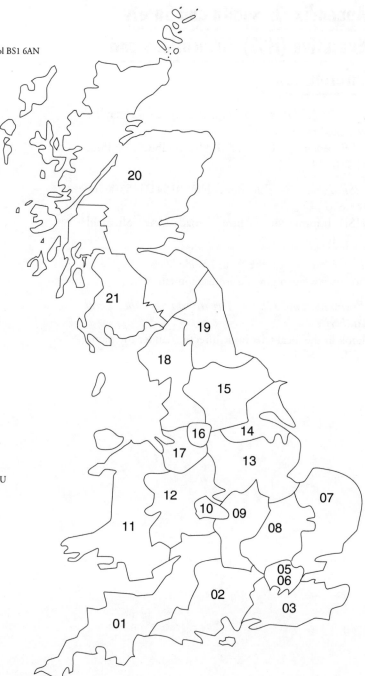

GLOSSARY

Adder In a computer it is a device which can form the sum of two or more numbers.

Address Information which identifies a particular location in the memory of a computer.

Aerial The part of a communication system from which energy is radiated or received.

ALGOL A symbolic language used to program computers in mathematical and engineering applications.

Alignment The adjustment of tuned circuits so that they respond in a desired way at a given frequency.

AM (amplitude modulation) A method of sending a message on a radio or light wave by varying the amplitude of the wave in response to the frequency of the message.

Amplifier A device whose output is a magnified function of its input.

Analogue signals Signals that respond to or produce a continuous range of values rather than specific values.

Analogue-to-digital converter A circuit designed to convert an analogue voltage into a binary code which can be read by a computer.

AND gate A building block in digital logic circuits.

Astable A circuit which can generate a continuous waveform with no trigger.

Atom The smallest particle of a chemical element that can exist alone or in combination with other atoms.

Attenuator A network designed to reduce the amplitude of a wave without distortion.

Audio frequency Any frequency at which the sound wave can normally be heard. The audio frequencies for most humans are those frequencies between 15 Hz and about 20 kHz.

Automation Any device or system that takes the place of humans in carrying out repetitive and boring jobs.

Band pass filter A filter that passes all frequencies between two specified frequencies.

Bandwidth The range of frequencies amplified or the range of frequencies passed by a filter.

BASIC (beginners all-purpose symbolic instruction code) An introductory high-level computer programming language.

Battery A power source made up of a number of individual cells.

Bias The current or voltage which is applied to part of a circuit to make the circuit function properly.

Binary number A number which can have just two values, 1 and 0.

Bipolar transistor A transistor that depends, for its operation, on both n-type and p-type semiconductors.

Bistable (also flip-flop) A circuit which has two stable outputs which can act as memories for data fed into its input.

Bit A unit of information content.

Boolean algebra A branch of symbolic logic in which logical operations are indicated by operators such as AND, OR and NOT signs.

Bourdon tube A pressure-measuring device made of a flexible tube formed into a C shape. Increasing pressure causes the C to straighten.

Buffer An isolating circuit used to avoid a reaction between a driver and driven circuit.

Bus One or more conductors used as a path for transmitting information from source to destination.

Byte A sequence of adjacent binary bits, usually 8.

Carrier wave A relatively high frequency wave which is suitable for transmission and modulation.

CATV (cable TV) A distribution of TV programmes by means of cables laid underground.

Cell A single source of electric potential.

Chip A small piece of silicon on which a complex miniaturised circuit, called an integrated circuit, is formed by photographic and chemical processes.

Circuit An electrical network in which there is at least one path which can be closed.

Closed loop control A control system which modifies its own behaviour according to feedback information e.g. constant speed control of an electric motor.

Closed loop gain The gain of an amplifier with feedback. Negative feedback reduces amplifier gain. Positive feedback increases amplifier gain.

CMOS (complementary metal oxide semiconductor logic) A logic family used especially in portable equipment.

Co-axial cable A cable formed from an inner and outer cylindrical conductor.

Code A system of symbols which represents information in a form which is convenient for a computer.

Colour code The values and tolerances of components such as resistors indicated by coloured bands.

Combinational logic A digital circuit, e.g. a NAND gate, that produces an output based on the combination of 0s and 1s presented to its input.

Communications The transmission of information by means of electromagnetic waves or by signals along conductors.

Comparator An electronic device, e.g. one based on an operational amplifier, that produces an output when the voltages of two input signals are different.

Computer A programmable device used for storing, retrieving and processing data.

CPU (central processor unit) The principal operating and controlling part of a computer, also known as its microprocessor.

Critically damped The degree of damping that provides the best compromise between the undamped response and the overdamped response.

CRO (cathode ray oscilloscope) A test and measurement instrument for showing the patterns of electrical waveforms and for measuring their frequency and other characteristics.

Cutoff frequency The 'corner frequency' of a filter. That frequency at which the signal level falls by 3 dB.

Data-handling Automatic or semi-automatic equipment which can collect, receive, transmit and store numerical data.

Decade counter A binary counter that counts up to a maximum count of ten before resetting to zero.

Decibel (dB) A unit used for comparing the strengths of two signals, such as the intensity of sound and the voltage gain of an amplifier.

Decoder A device that converts coded information, e.g. the binary code into a more readily understood code such as decimal.

Decoupling network A network designed to prevent an interaction between two electric circuits. These usually consist of RL or RC filters.

Demodulator A device for recovering information, e.g. music from a carrier wave.

DIAC Four-layer breakover device used to extend the range of control in a TRIAC circuit.

Dielectric The insulating layer between the conducting plates of a capacitor.

Difference amplifier An operational amplifier circuit that finds the difference between two input voltages.

Differential pressure flowmeter A device for measuring the flow of fluids. Depends on the drop in pressure created when a fluid flows past an obstruction or around a bend.

Differentiator An amplifier that performs the calculus operation of differentiation.

Diffusion The movement of electrons and/or holes from a region of high to low concentration.

Digital computer A system that uses gates, flip-flops, counters etc., to process information in digital form.

Digital-to-analogue converter A device that converts a digital signal into an equivalent analogue signal. DACs are widely used in computer systems for controlling the speed of motors, the brightness of lamps etc.

Digital voltmeter A voltmeter which displays the measured value as numbers composed of digits.

Doping The process of introducing minute amounts of material, the dopant, into a silicon crystal to produce n-type or p-type semiconductors in the making of transistors, integrated circuits and other devices.

Earth electrode A conductor driven into the earth and used to maintain conductors connected to it at earth potential.

Electric charge The quantity of electricity contained in or on a body, symbol Q, measured in coulombs, symbol C.

Electrolyte A substance which produces a conducting medium when dissolved in a suitable solvent.

Electrolytic capacitor A capacitor which is made from two metal plates separated by a very thin layer of aluminium oxide. Electrolytic capacitors offer a high capacitance in a small volume, but they are polarised and need connecting the right way round in a circuit.

Electromagnetic spectrum The family of radiations which all travel at the speed of light through a vacuum and include light, infrared and ultraviolet radiation.

Electron A small negatively charged particle which is one of the basic building blocks of all substances and forms a cloud round the nucleus of an atom.

Electronic ice A system of reference junction compensation used in thermocouple circuits. This is an electronic means of creating the thermocouple reference junction.

Electronics That branch of science and technology which is concerned with the study of the conduction of electricity in a vacuum and in semiconductors and with the application of devices using these phenomena.

emf (electromotive force) The electrical force generated by a cell or battery that makes electrons move through a circuit connected across the terminals of the battery.

Encoder Any device that converts information into a form suitable for transmission by electronic means.

Fan-in The number of logic gate outputs which can be connected to the input of another logic gate.

Fan-out The number of logic gate inputs which may be driven from a logic gate output.

Farad (F) The unit of electrical capacitance and equal to the charge stored in coulombs in a capacitor when the potential difference across its terminals is 1 V.

FAX (facsimile) The process of scanning fixed graphic material so that the image is converted into an electrical signal which may be used to produce a recorded likeness of the original.

Feedback The sending back to the input part of the output of a system in order to improve the performance of the system. There are two types of feedback, positive and negative.

Ferrite One of a class of magnetic materials which have a very low eddy-current loss, used for high-frequency circuit transformers and computer memories.

Ferroxcube A commercially available ferrite.

FET (field-effect transistor) A unipolar transistor that depends for its operation on either n-type or p-type semiconductor material.

Fibre optics The use of hair-thin transparent glass fibres to transmit information on a light beam that passes through the fibre by repeated internal reflections from the walls of the fibre.

Fidelity The quality or precision or the reproduction of sound.

Filter A circuit that passes only signals of a desired frequency or band of frequencies. May be high pass, low pass or band pass.

Flip-flop A device having two stable states, logic 0 or logic 1, and two input terminals corresponding to these states. The device will remain in either state until caused to change to the other by the application of an appropriate signal. In digital electronics, the bistable multivibrator circuit has earned the name 'flip-flop'.

Float switch Level-sensing limit switch, actuated by a float on the surface of a liquid.

Floppy disc A flexible disc, usually 5.25 inches (133 mm) in diameter, made of plastic and coated with a magnetic film on which computer data can be stored and erased.

FM (frequency modulation) A method of sending information by varying the frequency of a radio or light wave in response to the amplitude of the message being sent. For high-quality radio broadcasts, FM is preferable to AM since it is affected less by interference from electrical machinery and lightning.

Force A directed effort that changes the motion of a body.

Forward bias A voltage applied across a p–n junction which causes electrons to flow across the junction.

Forward breakover voltage The voltage between anode and cathode of an SCR at which forward bias conduction will begin.

Fourier analysis A mathematical method of determining the harmonic component of a complex wave.

Gain The ratio of increase in signal level between the input and output of an amplifier.

Gauge factor The ratio of change in resistance to the change in length of a strain gauge. Approximately 2 for a bonded foil strain gauge.

Gigabyte (GB) A quantity of computer data equal to one thousand million bytes.

Gigahertz (GHz) A frequency equal to one thousand million hertz (10^9 Hz)

Half-wave rectifier A diode, or circuit based on one or more diodes, which produces a direct current from alternating current by removing one half of the a.c. waveform.

Hardware Any mechanical or electronic equipment that makes up a system.

Heat sink A relatively large piece of metal that is placed in contact with a transistor or other component to help dissipate the heat generated within the component.

Henry (H) The unit of electrical inductance.

Hertz (Hz) The unit of frequency equal to the number of complete cycles per second of an alternating waveform.

High pass filter A filter that passes all frequencies above a specified frequency.

Hole A vacancy in the crystal structure of a semiconductor that is able to attract an electron. A p-type semiconductor contains an excess of holes.

Impedance (Z) The resistance of a circuit to alternating current.

Impurity An element such as boron that is added to silicon to produce a semiconductor with desirable electrical qualities.

Inductor An electrical component, usually in the from of a coil of wire. Inductors are used as 'chokes' to reduce the possibly damaging effects of sudden surges of current, and in tuned circuits.

Information technology (IT) The gathering, processing and circulation of information by combining the data-processing power of the computer with the message-sending capability of communications.

Input/output port The electrical 'window' on most computer systems that allows the computer to send data to and receive data from an external device.

Instrumentation amplifier A difference amplifier with very high input impedances at both inputs.

Insulator A material, e.g. glass, that does not allow electricity to pass through it.

Integrated circuit (IC) An often very complex electronic circuit which has resistors, transistors, capacitors and other components formed on a single silicon chip.

Integrator An amplifier circuit that performs the calculus function of integration.

Interface A circuit or device, e.g. a modem, that enables a computer to transfer data to and from its surroundings or between computers.

Inverting amplifier An amplifier whose output is 180 degrees out of phase with its input.

Ion An atom or group of atoms that has gained or lost one or more electrons, and which therefore carries a positive charge.

Isothermal block A connecting block used with thermocouples.

Jack A connecting device which is arranged for the insertion of a plug to which the wires of a circuit may be connected.

Joule The SI unit of energy.

Junction A region of contact between two dissimilar metals (as in a thermocouple) or two dissimilar conductors (as in a diode) which has useful electrical properties.

Kilobit One thousand bits. i.e. 0s and 1s of data.

Kilobyte One thousand bytes of data.

Kilohertz (kHz) A frequency equal to 1000 Hz.

Large-scale integration (LSI) The process of making integrated circuits with between 100 and 5000 logic gates on a single silicon chip.

Laser A device that produces an intense and narrow beam of light of almost one particular wavelength. The light from lasers is used in optical communications systems, compact disc players and video disc players. Laser is an acronym for Light Amplification by Stimulated Emission of Radiation.

LCD (liquid crystal display) A display that operates by controlling the reflected light from special liquid crystals, rather than by emitting light as in the light-emitting diode.

LDR (light-dependent resistor) A semiconductor device that has a resistance decreasing sharply with increasing light intensity. The LDR is used in light control and measurement systems, e.g. automatic street lights and cameras.

LED (light-emitting diode) A small semiconductor diode that emits light when current passes between its anode and cathode terminals. Red, green, yellow and blue LEDs are used in all types of display systems, e.g. hi-fi amplifiers.

Limit switch A switch that is arranged to be actuated by a workpiece.

Load The general name for a device e.g. an electric motor, that absorbs electrical energy.

Load cell A device for measuring weight. Weight resting on the device causes compression strain. Weight suspended from the device causes tensile strain. Strain is reported as a change in resistance by a coupled strain gauge.

Logic circuit An electronic circuit that carries out simple logic functions.

Logic diagram A circuit diagram showing how logic gates and other digital devices are connected together to produce a working circuit or system.

Logic gate A digital device e.g. an AND gate, that produces an output of logic 1 or 0 depending on the combination of 1s and 0s at its inputs.

Loudspeaker A device used to convert electrical energy into sound energy.

Low pass filter A filter that passes all frequencies below a specific frequency.

LVDT (linear variable differential transformer) A device used for position detection.

Machine code Instructions in the form of patterns of binary digits which enable a computer to carry out calculations.

Magnetic bubble memory (MBM) A device that stores data as a string of magnetic 'bubbles' in a thin film of magnetic material. The MBM can store a very large amount of data in small volume and is ideal for portable computer products such as word processors.

Magnetic reed switch A magnetically operated switch. Made of two or three magnetic leaves in a glass tube. Proximity of a magnet causes the switch to close.

Magnetic storage Magnetic tapes, floppy discs and magnetic bubble memories that store data as local changes in the strength of a magnetic field, and which can be recovered electrically.

Majority carrier The most abundant of the two charge carriers present in a conductor. The majority charge carriers in n-type material are electrons.

Man-machine interface Any hardware, e.g. a keyboard or mouse, that allows a person to exchange information with a computer or machine.

Mark-to-space ratio The ratio of the time that the waveform of a rectangular waveform is *high* to the time it is *low*.

Mass flowmeter A fluid-flow measuring device that measures the mass of the fluid instead of its velocity. Used when great accuracy is required.

Matrix A logical network in the form of a rectangular array of intersections.

Medium waves Radio waves having wavelengths in the range about 200 to 700 m, i.e. frequencies in the range 1.5 to 4.5 MHz.

Megabit A quantity of data equivalent to one million (10^6) bits.

Megabyte A quantity of binary data equal to one million (10^6) bytes. Floppy discs store approximately this amount of data.

Memory That part of a computer system used for storing data until it is needed. A microprocessor in a computer can locate and read each item of data by using an address.

Memory-mapped interface An interface system in which the input/output ports are addressed as memory locations.

Microcomputer A usually portable computer which can be programmed to perform a large number of functions quickly and relatively cheaply. Its main uses are in the home, school, laboratory and office.

Microelectronics The production and use of complex circuits on silicon chips.

Microfarad A unit of electrical capacitance equal to one millionth of a farad (10^{-6}F).

Micron (micrometre) A distance equal to one millionth of a metre. The micron is used for measuring the size and separation of components on silicon chips.

Microprocessor A complex integrated circuit manufactured on a single silicon chip. It is the 'heart' of a computer and can be programmed to perform a wide range of functions. A microprocessor is used in washing machines, cars, cookers, games and many other products.

Microswitch A small mechanically operated switch.

Microwaves Radio waves having wavelengths less than about 300 mm and used for straight line communications by British Telecom and others.

Minority carrier The least abundant of the two charge carriers present in a semiconductor. The minority charge carriers in n-type material are holes.

Modem (modulator/demodulator) A device for converting computer data in digital form into analogue signals for transmission down a telephone.

Modulator A circuit that puts a message on some form of carrier wave.

Monostable A circuit that produces a time delay when it is triggered, and then reverts back to its original, normally stable, state.

Mouse A small hand-operated device connected to a computer by a trailing wire, or by optical means, that makes a cursor move around the screen of a VDU to select operations and make decisions.

MSB (most significant bit) The left-hand binary digit in a digital word.

Multimeter An instrument for measuring current, potential difference and resistance, and used for testing and fault-finding in the design and use of electronic circuits.

Multiplexing A method of making a single communications channel carry several messages.

Multivibrator Any one of three basic types of two-stage transistor circuit in which the output of each stage is fed back to the input of the other stage using coupling capacitors and resistors, and causing the transistors to switch on and off rapidly. The multivibrator family includes the monostable, astable and bistable.

Negative feedback The feeding back to the input of a system a part of its output signal. Negative feedback reduces the overall gain of an amplifier but increases its bandwidth and stability.

Neutron A particle in the nucleus of an atom which has no electrical charge and a mass roughly equal to that of the proton.

Noise An undesirable electrical disturbance or interference.

Noise (white) Noise which is made up of a frequency spectrum (like white light).

Node A point of zero voltage or zero current on a conductor or the point in a radio wave where the amplitude is zero.

Nucleus The central and relatively small part of an atom that is made up of protons and neutrons.

Open loop control A control system in which no self-correcting action occurs as it does in a closed loop system.

Open loop gain The gain of an amplifier without feedback.

Operational amplifier (Op amp) A very high gain amplifier that produces an output voltage proportional to the difference between its two input voltages. Op amps are widely used in instrumentation and control systems.

Optical fibre A thin glass or plastic thread through which light travels without escaping from its surface.

Opto electronics A branch of electronics dealing with the interaction between light and electricity. Light-emitting diodes and liquid crystal displays are examples of opto electronic devices.

Oscillator A circuit or device, e.g. an audio frequency oscillator, that provides a sinusoidal or square wave voltage output at a chosen frequency. An astable multivibrator is one type of oscillator.

Package The plastic or ceramic material used to cover and protect an integrated circuit.

Parabolic reflector A hollow concave reflector.

Passband The range of frequencies passed by a filter.

Passive filter A filter made of passive components: resistors, capacitors and inductors.

PCB (printed circuit board) A thin board made of electrically insulating material (usually glass fibre) on which a network of copper tracks is formed to provide connections between components soldered to the tracks.

Period The time taken for a wave to make one complete oscillation. The period of the 50 Hz mains frequency is 0.02s.

Photodiode A light-sensitive diode that is operated in reverse bias. When light strikes the junction the diode goes into reverse breakdown and conducts. It is able to respond rapidly to changes of light.

Photoelectron An electron released from the surface of a metal by the action of light.

Photomask A transparent glass plate used in the manufacture of integrated circuits on a silicon chip.

Photon The smallest 'packet', or quantum, of light energy.

Photoresist A light-sensitive material that is spread over the surface of a silicon wafer from which silicon chips are made.

Photoresistor A transistor that responds to light and produces an amplified output signal. Like photodiodes, photoresistors respond rapidly to light changes and are used as sensors in optical communications systems.

Photovoltaic The property of responding to light with an electrical current. Photovoltaic cells are used in generating electricity from solar energy.

Picofarad (pF) An electrical capacitance equal to one millionth of a microfarad (10^{-12}F).

Piezoelectricity The electricity that crystals, such as quartz, produce when they are squeezed. Conversely, if a potential difference is applied across a piezoelectric crystal, it alters shape slightly. The piezoelectric effect is used in digital watches, hi-fi pick-ups and gas lighters.

Plasma A completely ionised gas at extremely high temperatures.

Port A place on a microcomputer to which peripherals can be connected to provide two-way communication between the computer and the outside world.

Positive displacement flowmeter A device that measures fluid flow by passing the fluid in measured increments. Usually accomplished by alternately filling and emptying a chamber.

Positive feedback The feeding back to the input of a system a part of its output signal. Positive feedback increases the overall gain of an amplifier and is used in an astable multivibrator.

Potential divider Two or more resistors connected in series through which current flows to produce potential differences dependent on the resistor values.

Potentiometer An electrical component, having three terminals, that provides an adjustable potential difference.

Power The rate of doing work. Measured in watts.

Preferred value Manufacturers' standardised component values used in resistor and capacitor values.

Programme A set of instructions used for the collation of data or for the solution of a problem.

Proton A particle that makes up the nucleus of an atom and has a positive charge equal in value to the negative charge of the electrons.

Pulse A short-lived variation of voltage or current in a circuit.

Q-factor The sharpness (or 'quality') of an electronic filter circuit, e.g. a tuned circuit, that enables it to accept or reject a particular frequency.

Quantum The smallest packet of radiant energy, e.g. a photon, that can be transmitted from place to place and described by Planck's quantum theory.

Quartz A crystalline form of silicon dioxide which has piezoelectric properties and can, therefore, be used as a pressure transducer and to provide a stable frequency in, for example, crystal clocks.

Qwerty keyboard A keyboard (e.g. a computer keyboard) that has its keys arranged in the same way as those of a standard typewriter, i.e. the first six letters of the top row spell 'QWERTY'.

Radiation Energy travelling in the form of electromagnetic waves.

Radio The use of electromagnetic waves to transmit or receive electrical signals without connecting wires.

RAM (random access memory) An integrated circuit that is used for the temporary storage of computer programs.

Rectifier A semiconductor diode that makes use of the one-way conducting properties of a p–n junction to convert a.c. to d.c.

Relay A magnetically operated switch that enables a small current to control a much larger current in a separate circuit.

Resistance The opposition offered by a component to the passage of electricity.

Resonance The build-up of large amplitude oscillations in a tuned circuit.

Response time The time required for a system to return to normal following a disturbance.

Reverse bias A voltage applied across a p–n junction (e.g. a diode) which prevents the flow of electrons across the junction.

Rheostat An adjustable resistor.

rms (root mean square) value The value of an alternating current which has the same heating effect as a steady d.c. current. 230 V is the rms value of the mains voltage.

Robot A computer-controlled device that can be programmed to perform repetitive tasks such as paintspraying, welding and machining of parts.

ROM (read-only memory) An integrated circuit that is used for holding data permanently, e.g. for storing the language and graphics symbols used by a computer.

Schmitt trigger A snap-action electronic switch which is widely used to 'sharpen up' slowly changing waveforms.

SCR (silicon-controlled rectifier) A four-layer semiconductor device used in switching circuits. Also known as a thyristor.

Semiconductor A solid material that is a better electrical conductor than an insulator but not such a good conductor as a metal. Diodes, transistors and integrated circuits are based on n-type and p-type semiconductors.

Sensor Any device which produces an electrical signal indicating a change in its surroundings.

Sequential logic A digital circuit that can store information. Sequential logic circuits are based on flip-flops and are the basis of counters and computer memories.

Servosystem An electromechanical system which uses sensors to control and monitor precisely the movement of something.

Short waves Radio waves that have wavelengths between about 2.5 MHz and 15 MHz and which are mainly used for amateur and long-range communications.

Silicon An abundant non-metallic element used for making diodes and transistors. Silicon is doped with small amounts of impurities such as boron and phosphorus to make n-type and p-type semiconductors.

Silicon chip A small piece of silicon on which a complex miniaturised circuit (called an integrated circuit) is formed by photographic and chemical processes.

Small-scale integration (SSI) The process of making integrated circuits.

Software Instructions or programs stored in a computer system.

Solder An alloy of tin and lead that has a low melting point and is used for making electrical connections between components on a circuit board.

Solenoid A coil of copper wire in which an iron rod moves by the magnetic field produced when a current flows through the coil.

Stepping motor An electric motor with a shaft that rotates one step at a time. Stepping (or stepper) motors are used for the precise positioning of robot arms.

Strain Change in dimension of a material when force is applied.

Strain gauge A device used to measure strain. The change in electrical resistance of the strain gauge is a measure of the strain.

Summing amplifier An amplifier whose output is proportional to the sum of two or more input signals, or an amplifier used to add (sum) its input voltages.

System All the parts which make up a working whole.

Tachogenerator A device used to measure motor speed. The output is a voltage or a frequency that is proportional to motor speed.

Telex An audio frequency teleprinter system provided by the Post Office for use over telephone lines.

Tensile strain Strain caused by force pulling on a member.

Thermistor A semiconductor temperature sensor.

Thermocouple A temperature-sensing device whose output is a current or voltage which is proportional to the difference between the temperatures at two junctions of dissimilar metals.

Thermopile A system of several thermocouples in a series-aiding configuration. This configuration increases the sensitivity of the thermocouple.

Thyristor A half-wave semiconductor switching device used for motor speed control and lamp dimming. Also known as a silicon-controlled rectifier (SCR).

Time constant The time taken for the voltage across a capacitor to rise to 63% of its final voltage when it charges through a resistor connected in series with it.

Torque Twisting or rotary force, such as that delivered by a motor shaft.

Transducer A device which converts mechanical or physical quantities into electrical quantities, or a device which converts electrical quantities into physical quantities.

Transformer An electromagnetic device for converting alternating current from one voltage to another.

Transistor A semiconductor device which has three terminals and is used for switching and amplification.

Transmitter A device or equipment which converts audio or video signals into modulated radio frequency signals which are then sent (transmitted) by electromagnetic waves.

TRIAC A full-wave semiconductor switching device used for motor speed control and lamp dimming.

Truth table A list of 0s and 1s that shows how a digital logic circuit responds to all possible combinations of binary input signals.

TTL (transistor-transistor logic) The most common type of IC logic in use today.

Tuned circuit A circuit which contains an inductor and a capacitor and can be tuned to receive particular radio signals.

Tweeter A loudspeaker used to reproduce the higher audio frequencies (above 5 kHz).

UHF (ultra-high frequency) Radio waves that have frequencies in the range 500 MHz to 30 000 MHz and are used for TV broadcasts.

UJT (uni-junction transistor) A type of transistor used as a relaxation oscillator in SCR control circuits.

Unipolar transistor A transistor that depends for its operation on either n-type or p-type semiconductor materials as in a field-effect transistor.

Ultrasonic waves Sound waves inaudible to the human ear that have frequencies above about 20 kHz.

Ultraviolet Radiation having wavelengths between the visible violet and the X-ray region of the electromagnetic spectrum.

VDU (video display unit) An input/output device comprising a screen and sometimes a keyboard that enables a person to communicate with a computer.

Velocity flowmeter A device that measures fluid flow directly. The most common is the turbine flowmeter.

VHF (very high frequency) Radio waves that have frequencies in the range 30 MHz to 300 MHz and are used for high-quality radio broadcasts (FM) and TV transmission.

Viewdata An information service that enables telephone subscribers to access a wide range of information held in a database and which is displayed on a TV set coupled to the telephone line by a modem.

VMOS (vertical metal-oxide semiconductor) A type of field-effect transistor. It is a small high-power fast-acting transistor used in audio amplifiers and power switching circuits.

Voltage difference amplifier An amplifier whose output is proportional to the difference between two input voltages.

Wafer A thin disc cut from a single crystal of silicon on which hundreds of integrated circuits are made before being cut up into individual ICs for packaging.

Waveform The shape of an electrical signal, e.g. a sinusoidal waveform.

Wavelength The distance between one point on a wave and the next corresponding point.

Wheatstone bridge A network of resistors used to measure very small changes in resistance.

Woofer A loudspeaker used to reproduce the lower audio frequencies.

Word A pattern of bits (i.e. 1s and 0s) that is handled as a single unit of information in digital systems; e.g. a byte is an 8-bit word.

Wordprocessor A computerised typewriter that allows written material to be generated, stored, edited, printed and transmitted.

X-axis deflection Horizontal deflection on the screen of a CRO, often used as the time base.

X-rays Penetrating electromagnetic radiation used in industry and in medicine for seeing below the surface of solid materials.

Y-axis deflection Vertical deflection on the screen of a CRO.

Zener diode A special semiconductor diode that is designed to conduct current in the reverse-bias direction at a particular reverse-bias voltage. Zener diodes are widely used to provide stabilised voltages in electronic circuits.

SOLUTIONS TO EXERCISES

CHAPTER 1

1:c 2:b 3:c 4:b 5:b 6:a 7:b 8:a 9:b 10:c 11:d 12:b 13:c 14:a 15:b 16:c 17–25: answers in text

CHAPTER 7

1: b
2: d
3: b
4: See Fig. 7.4.
5: See Fig. 7.6.
6: See Fig. 7.33.

X	Y	Z	F
0	0	0	1
0	1	1	0
1	0	1	0
1	1	1	0

Figure 7.33 Truth table for question number 6.

7: See Fig. 7.34. The output is high only when the input X is low and the input Y is high. For all other input combinations the output is low.

X	Y	Z	F
0	0	1	0
0	1	1	1
1	0	0	0
1	1	0	0

Figure 7.34 Truth table for question number 7.

8: See Fig. 7.35. The output is high for all input combinations except when input X is low and input Y is high.

X	Y	Z	F
0	0	1	1
0	1	0	0
1	0	1	1
1	1	0	1

Figure 7.35 Truth table for question number 8.

9: See Fig. 7.36. The output X is at logic 1 only when input P and input Q are at logic 1. For all other input combinations the output is logic 0.

P	Q	R	S	T	X
0	0	1	1	1	0
0	1	1	0	1	0
1	0	0	1	1	0
1	1	0	0	0	1

Figure 7.36 Truth table for question number 9.

10: See Fig. 7.37. The output T is logic 1 only when both inputs are at logic 1. For all other input combinations the output is logic 0.

P	Q	R	S	T
0	0	0	0	0
0	1	0	1	0
1	0	0	1	0
1	1	1	1	1

Figure 7.37 Truth table for question number 10.

11: See Fig. 7.38. The output E is logic 1 for all input combinations except when input A and B are both logic 0.

A	B	C	D	E
0	0	0	0	0
0	1	0	1	1
1	0	0	1	1
1	1	1	0	1

Figure 7.38 Truth table for question number 11.

12: See Fig. 7.39. The output is high when both inputs are the same.

A	B	C	D	E	F	G
0	0	1	1	1	0	1
0	1	1	0	0	0	0
1	0	0	1	0	0	0
1	1	0	0	0	1	1

Figure 7.39 Truth table for question number 12.

13: See Fig. 7.40. An output is only available at F when keys A and B are off (both at logic 0) and key C is on.

A	B	C	D	E	F
0	0	0	1	1	0
0	0	1	1	1	1
0	1	0	1	0	0
0	1	1	1	0	0
1	0	0	0	1	0
1	0	1	0	1	0
1	1	0	0	0	0
1	1	1	0	0	0

Figure 7.40 Truth table for question number 13.

CHAPTER 8

1:c 2:c 3:d 4:c 5:b 6:c 7:d 8:b 9:a 10:d 11:a 12:d 13:a 14:c 15:c 16:a 17:b 18:a 19:d 20:b 21:b 22:b 23:a 24:b 25:b 26:c 27:b 28:(a) 295 V (b) 240 V (c) 217 V 29–35:Answers in text. 36:5.03 kHz. 37:Answers in text. 38:25:1

INDEX

Abbreviations in electronics, 209
Acceleration, 140
Accident reports, 19
Accidents at work, 23
Accuracy, 93
A.c. supplies, 125
Airways, 18
Alarm call points, 180
Alternating current theory, 125
Ammeter, 125
Amplification, 83
Amplitude modulation (AM), 90
AM receiver, 155
AM transmitter, 155
Analogue electronics, 104
Analogue instruments, 94
Astable multivibrator, 92
Astra satellite, 168
Atoms and electronics, 114
Audio connectors, 55
Automatic fire sensors, 181
Average value, 126
Avo meter, 96
Azimuth angle, 168

Bandwidth, 85, 88
Batteries:
    charging information, 215
    discharging information, 216
    power rating, 214
    supplies, 42
Bending strain, 188
Binary numbers, 104
Bipolar transistor, 64
Bleeding, 16
Block diagrams, 160
BNC connectors, 174

Bourdon tube, 193
Breadboards, 54
Breathing stopped, 18
Bridge rectifiers, 222
British Standards, 5
Broken bones, 17
BSI kite mark, 6
BSI safety mark, 6
Buffers, 105
Burns, 16
Burst trigger control, 79
Bypass diode, 81

Cable connectors, 55
Cable television, 172
Capacitive reactance, 127
Capacitors, 34
    charging, 117
    colour code, 38
    in combination, 123
    rating, 75
    testing, 39
Cardiac arrest, 18
Cartridge fuse, 8
Cathode ray tube, 97
CCTV (closed circuit television), 179
CD-ROM, 160
CE safety mark, 6
Charging capacitors, 91
Chemical effect, 116
Choking, 17
Circuit boards, 51
Circuit breaker, 8
Circuit diagrams, 161
Clean supplies, 173
Closed loop control, 150
CMOS digital logic, 110

Codes of practice, 5
Colour, definition of, 145
Colour mixing, 146
Communication systems, 163
Competent person, 11, 102
Component information, 44
Computers:
    networks, 173
    supplies, 172
    systems, 159
Conduction, 143
Conductor, 114
Conservation of energy, 142
Control of Substances Hazardous to Health
    Regulations (COSHH) 1988, 2
Convection, 142
Copper strip board, 53
Coupling capacitors, 35
CPU, 160
CRO, 87, 97, 156
C–R time constant, 92, 117
Current gain, 66

Decibel, 4
D-connectors, 175
Decoupling capacitors, 35
Demodulator, 155
Desoldering, 50
Diac, 70, 82
Digital clock, 158
Digital electronics, 104
Digital meters, 94
DIL packages, 68
DIN connectors, 55
Diode, 59, 75, 217
Discrete component, 65
Display screen equipment, 25
Doping, 58
Dot/cross notation, 135
Dry joints, 48
DVD, 160

Earth bond test, 102
Earth leakage test, 102
Eddy current loss, 138
Electrical isolation, 9
Electrical machines, 136
Electrical units and symbols, 210
Electricity at Work Regulations, 5
Electric shock, 18
Electrolysis, 116

Electrolytic capacitors, 36
Electromagnetic relay, 41
Electromotive force, 114
Electronic symbol, 30
Electrostatics, 116
Emergency lighting, 184
Energy, 124, 142
Ergonomics, 23
Error, 93
European Standards, 6
Exclusive OR gate, 103
Exponential curve, 117

Faraday's law, 135
Fault finding, 56
Feedback effects, 152
Ferrite core, 138
Fibre optics, 164
Filters, 86
Fire:
    categories of, 20
    control, 20
    extinguishers, 21
Fire alarms:
    design considerations, 184
    sounders, 183
    system, 180
First aid, 15
    boxes, 17
First aider, 15
Flash test, 102
Float switch, 198
Fluid flow measurement, 198
Fluid level measurement, 197
Flux cored solder, 48
Flywheel diode, 81
FM receiver, 156
FM transmitter, 156
Force, 140
Force on a conductor, 136
Forward bias, 60
Frequency measurement, 98
Frequency modulation (FM), 90
Frequency response, 86, 88
Fuel, 20
Full-wave rectification, 73
Fundamental units, 113
Fuse, 8

Gain, 83
Generator, 137

Geostationary orbit, 167
Germanium, 58
Graphical symbols, 30

Half-wave rectification, 73
Hand tools, 46
Harmonics, 89
Hazardous substances, 2
Health and Safety at Work Act, 1
Heat, 21
Heat detectors, 182
Heat dissipation, 143
Heating effect, 116
Heat sink, 51, 87
Heat transfer, 142
High impedance, 93, 127
High pass filter, 86
HSE, 1, 4
HSE publication information, 237
Hysteresis loss, 138

IC, 67
IEE Regulations, 5
Impedance, 127
Inductive reactance, 127
Inductors, 39
Industrial sensors, 202
Inspection test, 100
Instability, 151
Instrument calibration, 93
Instrument connections, 125
Instrument errors, 93
Insulator, 114
Insulation test, 100
Integrated circuits, 67
Interface, 160
Intruder alarms:
   controls, 179
   design considerations, 179
   sounders, 179
   systems, 177
Iron core, 39
Isolation of supplies, 9, 100

Jack connectors, 56

Laminated iron core, 138
LANs (local area networks), 173
LDR (light-dependent resistor), 63
LED (light-emitting diode), 61
Lenses, 146

Light, 145
Light dependant resistor, 63
Light emitting diode, 61
Limit switches, 203
Live testing, 11
Logic at work, 111
Logic families, 109
Logic gates, 103
Logic networks, 107
Logic symbols, 232
Low pass filter, 75, 86

Magnetic effect, 116
Magnetic pick up, 200
Manual handling, 22
Mark to space ratio, 89
Martindale testing device, 9, 10
Mass, 140
Matching transistors, 85
Matrix board, 51
Maximum value, 126
MCB (miniature circuit breakers), 8
Measuring frequency, 98
Mechanics, 139
Microphone, 195
Microswitches, 41
Microwave telephones, 164
MilSpec symbols, 111
Mobile telephones, 164
Motor/generator principle, 136
Motor speed control, 150
Multimeter, 95
Multivibrator, 92

NAND gate, 104
Negative feedback, 151
Newton, 145
Noise, 4
Noise suppression, 173
NOR gate, 104
NOT gate, 104
N-type material, 58

Obtaining electronic information, 209
Ohm, George Simon, 114
Ohm meter, 95
Ohm's law, 114
Open loop control, 150
Operational amplifier, 84
Operator errors, 94
Optical fibres, 164

Optics, 164
Opto electronics, 207
Opto isolator, 208
OR gate, 103
Overcurrent protection, 8. 42
Oxygen, 20

Parallel capacitors, 123
Parallel resistors, 119
Pascals, 4
Passive detectors, 117
PAT testing, 100
PCB, 52
Personal hygiene, 23
Personal protection, 3, 4
P.f. (power factor) improvement, 129
Phase angle, 129
Phase control, 79
Phasor diagram, 128
Photo-electric detectors, 207
Photodiode, 65
Piezoelectric transducer, 192
PIR detectors, 176
Plug top, 101
Portable appliance testing, 100
Positional reference system, 52
Positive feedback, 151
Potential difference, 114
Power, 123, 141
Power amplifier, 84
Power control, 79
Power factor, 129
Power supplies, 99
PPE, 3
Preferred values, 34
Pressure measurement, 192
Pressure or stress, 140
Pressure pad, 178
Printed circuits, 52
Protoboard, 54
Proving unit, 10
Proximity switches, 177
PSU, 99
P-type material, 58
Pulse counter, 200
Pulse modulation, 90

Quartz crystal, 192

Radiation, 142
Radio transmission, 166

RAM, 160
RCD, 9
Reactance, 127
Record keeping, 103
Rectification, 76
Reed switches, 41
Reflection, 146
Refraction, 146
Regulation, 76
Relay, 41, 207
Resistors, 30
Resistivity, 115
Resistor colour code, 32, 34
Resistors in combination, 121
Resonance, 134
Reverse bias, 60
Ribbon cable, 55
Ribbon connectors, 55
RIDDOR, 19
Risk and hazard, 11
Risk assessment, 26
Rms value, 126
ROM, 160

Safety documentation, 2
Safety equipment, 3
Safety precautions, 45
Safety signs, 3,14, 15, 24
Satellite communications, 165
Satellite dish installation, 168
Satellite dish problems, 172
Satellite locations, 171
Satellite regulations, 172
Satellite television, 168
Sawtooth waveform, 91
Screw rule, 135
S-DeC boards, 54
Secure isolation, 11
Security lighting, 176
Security systems, 153, 176
Semiconductors, 58
Sensors, 186
Series capacitors, 123
Series resistors, 119
Series resonance, 134
Serious injuries, 17
Signal distortion, 87
Signal generator, 87, 99
Signal modulation, 90
Silicon, 58
Sinusoidal waveform, 125

INDEX

SI units, 113
Sky television, 168
Smoke alarms, 181
Smoothing, 73
Snubber network, 82
Soldering gun, 46
Soldering iron, 46
Soldering techniques, 47
Sound, 144
Space heating control, 154
Speed, 140
Speed of light/sound, 145
Stabilised power supply, 76
Statutory regulations, 5, 45
Stepper motor, 201
Strain gauges, 187
Strain gauges, specifications, 236
Strain measurement, 186
Stripboard, 552
Stroboscope, 199
Supplementary diagrams, 161
Supplies, 42
Switches, 40
Symbols used in electronics, 30
Systems, 149
System X, 164

Tachometer, 199
Tape recorder, 157
Telephone circuits, 163
Telephones, mobile, 164
Telephone systems, 163
Temperature and heat, 142
Temperature compensation, 188
Temperature, effects of change, 143
Temperature measurement, 194
Temporary circuit building, 54
Tensile strength, 189
Testing capacitors, 39
Testing resistors, 34
Test instruments, 93
Test probes, 11
Thermistor, 64, 196
Thermocouple, 194

Thermocouple, colour code, 235
Three effects of electric current, 116
Three-phase power control, 82
Thyristor operation, 68
Thyristor testing, 68
Time constant, 118
Touch voltage, 6
Transducer, 152, 186
Transformer construction, 138
Transformer losses, 138
Transformer matching, 138
Transformer principle, 136
Transistor, 64
Transistor pin connection, 223
Transistor switching, 91
Transistor testing, 66
Triac, 69
Truth table, 103
TTL digital logic, 110

Ultrasonic detectors, 178
Unidirectional supply, 125
Uninterruptable supplies, 9
Units, 113
UPS (uninterrupted power supplies), 9

VDU hazards, 24
VDU operator, 26
Velocity, 140
Veroboard, 52
Visual inspection, 101
Voltage divider, 71
Voltage indicator, 9, 100
Voltage regulators, 77, 227, 229
Voltmeter, 125

Wattmeter, 125
Wire wrapping, 53
Wiring diagrams, 161
Wiring regulations, 5
Wheatstone bridge, 187
Work done, 141

Zener diode, 59, 218